Jan Harmsen and Maarten Verkerk
**Process Intensification**

## Also of Interest

*Product and Process Design.*
*Driving Innovation*
Harmsen, deHaan, Swinkels, 2018
ISBN 978-3-11-046772-7, e-ISBN 978-3-11-046774-1

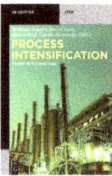

*Process Intensification.*
*Design Methodologies*
Gómez-Castro, Segovia-Hernández (Eds.), 2019
ISBN 978-3-11-056237-8, e-ISBN 978-3-11-056255-2

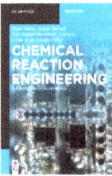

*Chemical Reaction Engineering.*
*A Computer-Aided Approach*
Salmi, Wärnå, Hernández Carucci, de Araújo Filho, 2020
ISBN 978-3-11-061145-8, e-ISBN 978-3-11-061160-1

*Process Engineering.*
*Addressing the Gap between Study and Chemical Industry*
Kleiber, 2020
ISBN978-3-11-065764-7, e-ISBN 978-3-11-065768-5

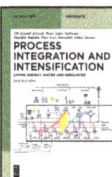

*Sustainable Process Integration and Intensification.*
*Saving Energy, Water and Resources*
Klemeš, Varbanov, Wan Alwi, Manan, 2018
ISBN978-3-11-053535-8, e-ISBN 978-3-11-053536-5

Jan Harmsen and Maarten Verkerk

# Process Intensification

——

Breakthrough in Design, Industrial Innovation Practices, and Education

**DE GRUYTER**

**Authors**
**Ir. Jan Harmsen**
Harmsen Consultancy B.V
HoofdwegZuid 18
2912 ED Nieuwerkerk a/d IJssel
Netherlands
info@harmsenconsultancy.nl

**Prof. Dr. Maarten Verkerk**
Slot Aldeborglaan 26
6432JM Hoensbroek
Netherlands
maarten.verkerk@home.nl

ISBN 978-3-11-065734-0
e-ISBN (PDF) 978-3-11-065735-7
e-ISBN (EPUB) 978-3-11-065752-4

Library of Congress Control Number: 2020935458

**Bibliographic information published by the Deutsche Nationalbibliothek**
The Deutsche Nationalbibliotheklists this publication in the Deutsche Nationalbibliografie;
detailed bibliographic data are available on the Internet at http://dnb.dnb.de.

© 2020 Walter de Gruyter GmbH, Berlin/Boston
Cover image: Ivan Bajic/iStock /Getty Images Plus
Printing and binding: CPI books GmbH, Leck

www.degruyter.com

# Preface

We, the authors, have made a very interesting journey together in writing this book. When we started, we had no idea where the journey would bring us. Each of us had some idea about the final destination. However, it was not the same destination. On the way, we met a couple of engineers who shared their experiences about process intensification with us. They also changed the route of our journey. We never dreamt of the final result of this book on process intensification design, industrial innovation practices, and education. We also did not expect such great cooperation from case writers from various process industry branches.

We would like to acknowledge all contributors to this book. First of all, the co-authors who wrote complete chapters of this book: Gerhard Muggen and his employees of BTG-BTL for providing the case on biomass pyrolysis to stable fuel oil. Despite being very busy with new projects, they produced their wonderful story; Jeff Siirola for his additions on values of engineers, interests of stakeholders, and beliefs of society to the Eastman case; Tony Kiss and Vladimir Maleta for their cyclic distillation case and their extra effort to get many details in the VIB perspectives of this case; Tuomas Koiranen for writing his pharma case chapter and appendix and always being willing to add and improve his contributions; Michiel van der Stelt and Jeroen van Gestel for providing the chapter on education and for their contributions to the OMEGA case chapter to make it more interesting for students.

Second, we would like to thank the reviewers of chapters for their comments. Frans Visscher reviewed Chapters 2, 6–11, and 13–15. Ger van der Horst reviewed Chapters 1, 3, 4, 5, and 19. Aris de Rijke reviewed Chapters 3 and 4. They provided a lot of very useful comments that greatly improved these chapters.

We would also like to thank our wives for putting up with husbands often totally focused on this book. Without their support the book would not have been completed in time.

https://doi.org/10.1515/9783110657357-202

# Contents

## Part A: **Theory**

## Part B: Application of theory: guidelines and methods

## Part C: **Industrial practice cases**

## Part D: Educating PI – academic and industrial

# Authors' Biographies

**ir. Jan Harmsen**
Harmsen Consultancy BV
Hoofdweg Zuid 18, 2912ED, Nieuwerkerk a/d IJssel, Netherlands
E-mail: jan@harmsenconsultancy.nl

Jan Harmsen is currently an independent consultant for sustainable process innovation. He provides advice and courses to industry and academia. After his graduation in chemical technology at Twente University, in 1977, he joined Shell. There he held professional positions in process research, process development, reaction engineering, process concept design, process implementation at manufacturing sites in Rotterdam and South Korea, and finally a research position in process intensification till 2010. He was part-time Hoogewerff-professor Sustainable Chemical Technology, first at Delft University of Technology, in 1997, and later at Groningen University till 2013.

He is coauthor of the following books: Jan Harmsen, *Industrial Process Scale-Up – A Practical Innovation Guide from Idea to Commercial Implementation*, 2nd revised edition, Elsevier, 2019; Jan Harmsen, et al., *Product and Process Design Driving Innovation*, De Gruyter, 2018; Gerald Jonker and Jan Harmsen, *Engineering for Sustainability: A Practical Guide for Sustainable Design*, Elsevier 2012; Jan Harmsen, Joseph Powell (eds.), *Sustainable Development in the Process Industries: Cases and Impact*, John Wiley & Sons, 2010.

**Dr. Maarten Johannes Verkerk**
Slot Aldeborglaan 26, 6432JM, Hoensbroek, Netherlands
E-mail: maarten.verkerk@home.nl

Maarten Verkerk is an independent consultant in the field of innovation, sustainability, and meaning of life. He retired in December 2019 as an affiliate professor in Christian philosophy and ethics at the Maastricht University and Technical University of Eindhoven. Presently, he is a member of the Senate.

Maarten Verkerk (1953) was born in Ruwiel, the Netherlands. He is married to Nienke Vegter, and they have four children. Maarten studied chemistry, theoretical physics, and philosophy at the University of Utrecht. In 1982 he got his PhD in material science at the Technical University Twente in Enschede with the thesis "Electrical conductivity and interface properties of oxygen ion conducting materials." In 2004, he defended a second thesis on the crossing point of organization science, technology, and philosophy: *Trust and Power on the Shop Floor – An Ethnographical, Ethical and Philosophical Study on Responsible Behaviour in Industrial Organisations.*

After his first PhD, Maarten worked as a senior researcher at the Nat. Lab. of Philips in Eindhoven. Then, he worked for more than 15 years in a number of management functions in the industrial sector of Philips in The Netherlands, Germany, and Taiwan. Later on, he worked for 5 years as a director of a psychiatric hospital in Maastricht, the Netherlands. From 2008 to 2017, he was chairman of the board of VitaValley, an innovation network in healthcare. From 2004 to 2018, he was an extraordinary professor of Christian philosophy at the Technical University of Eindhoven, and from 2008 to 2019 at the Maastricht University.

https://doi.org/10.1515/9783110657357-204

Maarten has published on materials science, organization science, movies, innovation science, philosophy, philosophy of technology, ethics of technology, sustainability, end-of-life issues, women and church, and politics.

### Gerhard Muggen

Gerhard Muggen is managing director and shareholder of BTG Bioliquids BV (BTG-BTL). As managing director, he is responsible for the strategy, worldwide marketing and sales, and roll out of the BTG-BTL pyrolysis technology. Gerhard has been working for Stork Thermeq for 8 years as Sales manager and the last four years as Vice President Marketing and Sales. In the last job, he was responsible for the international marketing and sales department and was directly involved in marketing and sales in China, India and Russia. Gerhard was also responsible for organizing and initiating new developments and marketing new technologies.

### BTG Biomass Technology Group
### Prof. Dr. Jeffrey J. Siirola

Jeff Siirola retired in 2011 as a technology fellow at Eastman Chemical Company in Kingsport Tennessee, where he had been for more than 39 years. Following retirement, for the next 6 years he held half-time positions as Distinguished Service Professor of Sustainable Energy Systems at Carnegie Mellon University and Professor of Engineering Practice at Purdue University, teaching chemical process design, process synthesis, and industrial chemical technology. Starting 2017 Fall Semester, he now teaches full time at Purdue but maintains a special faculty appointment at CMU. Siirola received BS in chemical engineering from the University of Utah in 1967 and PhD in chemical engineering from the University of Wisconsin-Madison in 1970. His areas of interest include chemical process synthesis, computer-aided process systems engineering, design theory and methodology, technology assessment, resource conservation, sustainable development, carbon management, and chemical engineering education.

Siirola is a member of the National Academy of Engineering and served as president of the American Institute of Chemical Engineers.

### Prof. Dr. ir. Anton A. Kiss

University of Manchester, Department of Chemical Engineering & Analytical Science
Sackville Street, Manchester, M13 9PL, United Kingdom. E-mail: TonyKiss@gmail.com

Tony Kiss is a professor and chairs in chemical engineering at the University of Manchester, and a *Royal Society Wolfson Research Merit Award* holder. He holds a Chemical Engineer degree from Babes-Bolyai University of Cluj-Napoca and PhD from University of Amsterdam (UvA). He was also postdoctoral research fellow at TU Delft and University of Amsterdam. Tony is chartered fellow of *IChemE*, senior member of *AIChE*, and research fellow of *The Royal Society*, with 20+ years of academic and industrial experience. Previously, he worked for over a decade as senior project manager and RD&I specialist at AkzoNobel Chemicals (now rebranded as Nouryon). Besides his senior industrial role, he was also appointed as part-time professor of separation technology at University of Twente.

During the past decades, he has carried out many research and industrial projects, supervised graduation projects, published over 15 textbooks and book chapters, and over 100 scientific articles in highly ranked journals. His research focus is on process systems engineering, process intensification, and separation technology. Tony is also member of scientific committees (e.g., ESCAPE, D&A, and IPIC conferences), advisory and editorial boards (e.g., *J. Chem. Tech. Biot.* and *Chem. Eng. Res. Design*).

For his pioneering research work, he has received the *Hoogewerff Jongerenprijs* (a prestigious award recognizing the most promising young scientist in the Netherlands), the *AkzoNobel Innovation Excellence Award* (for the most successful industrial innovation), the *Royal Society Wolfson Research Merit Award* (Manchester, UK), the *Pirkey Distinguished Lecturer* in Chemical Engineering (University of Texas at Austin, USA), and the *CHEMCON Distinguished Speaker Award* for Innovators and Science Leaders (IIChE, New Delhi). More information is available at: www.tonykiss.com

**Dr. Vladimir N. Maleta**
Maleta Cyclic Distillation LLC OÜ, Parnu mnt 130–38, 11317 Tallinn, Estonia

Vladimir Maleta is an entrepreneur who successfully developed and implemented new cyclic distillation technology in industry. He holds MSc in mechanical engineering from the Kiev Institute of Food Industry and PhD in engineering from the National University of Food Technologies, Ukraine. His PhD was focused on the mass transfer intensification of cyclic distillation systems for food industry. As the collapse of the Soviet Union and various political movements temporarily stopped his scientific activity, he worked at the Institute of Bioenergy Crops and Sugar Beet and in the banking sector. Then, he started his own company (Maleta Cyclic Distillation LLC) and began implementing scientific developments in the industry. More information is available at: www.maletacd.com

**Prof. Dr. Tuomas Koiranen**
Professor, Chemical process systems engineering
LUT School of Engineering Science
LUT University
Yliopistonkatu 34, FI – 53850 Lappeenranta, Finland
P.O. BOX 20, FI – 53851 Lappeenranta
tuomas.koiranen@lut.fi

Tuomas Koiranen did his MSc in chemical engineering on crystallization scale-up. He obtained his doctorate degree in 1998 in artificial intelligence applications in chemical process industries. He worked a couple of years in the Helsinki University of Technology as a post-doc, where his subjects were multi-phase modeling in computational fluid dynamics and crystallization modeling. He has 12 years of experience in pharmaceutical industrial research design as a process chemist and as a senior researcher. His expertise in industrial sector was process development, and crystal engineering. He started the tenure track with professorship in "Fluid Dynamics and Chemical Process Applications" in 2013 at LUT University in Lappeenranta, Finland. He has over 70 peer-reviewed articles and conference publications. He has guided five doctorate theses and over 50 MSc theses, most of them in chemical industries and in biorefinery sector. His current chair is professor in "Chemical Process Systems engineering." Chemical process intensification is closely connected both to research and teaching activities. The research topics are related to experimentation and modeling in chemical processes, crystallization, reactive extraction modeling, carbon negative/neutral process development and chemical syntheses in carbon dioxide chemistry. He belongs to the Working Party of Process Intensification in European Federation of Chemical Engineering. He has worked as a reviewer in chemical engineering and processing: process intensification; chemical engineering science; chemical engineering communications, and waste management, to name a few.

**Jeroen A.M. van Gestel**
Chemical Engineering Group, Utrecht University of Applied Sciences
Heidelberglaan 7, 3584 CS Utrecht, the Netherlands

Jeroen van Gestel has worked as a Lecturer of Chemical Engineering for over ten years. He holds an MSc degree in Chemical Engineering from *Eindhoven University of Technology* in the Netherlands and a PhD in Theoretical Polymer Physics from the same university. After several years as a researcher at various universities, Jeroen rediscovered his passion for education. Besides teaching practical and theoretical courses and coaching (groups of) students, he takes an active role in curriculum redesign and maintains a close connection with the workfield in his role as internship co-ordinator.

**Michiel J.C. van der Stelt**
Chemical Engineering Group, Utrecht University of Applied Sciences
Heidelberglaan 7, 3584 CS Utrecht, the Netherlands

Michiel van der Stelt has worked as a Lecturer of Chemical Engineering for over three years. He holds an MSc degree in Chemical Engineering from *Eindhoven University of Technology* in the Netherlands and a PhD in Chemical Engineering from the same university. Michiel worked several years in industry working on process improvement and development. Besides teaching practical and theoretical courses and coaching (groups of) students, he takes an active role in the coordination of the minor program Process Development in the Chemical Industry and maintains a close connection with the industry in coordinating partnerships and collaborative projects.

# 1 Introduction

## 1.1 Objectives of this book

The objectives of this book are threefold. The first objective is to provide engineers in the process industries design, methods for process intensification in combination with stage-gate innovation steps that include modeling and experimental validation, so that in the end, radically improved processes are implemented.

The second objective is to provide managers in the process industries, a practice perspective to understand the different cultures of research & development (R&D), operations (OPS), manufacturing, and marketing & sales (M&S). These practice perspectives support managers to facilitate cooperation between these different departments, so that these radical process innovation projects become commercial-scale successes – as they are accepted by the internal and the external stakeholders, and strongly contribute to the realization of the sustainable development goals [1]. The book is not only for large companies but also for start-up companies. To that end, it contains a start-up company case describing all hurdles and how they were overcome.

The third objective is to provide implemented industrial cases from various process industries, notably from chemicals, food, pharma and biofuels, and exercises for process engineering education programs such as in chemical engineering, food processing, pharmaceuticals, and minerals processing, so that the book can be used as a textbook for academic and industrial training.

The book is therefore not just a description of process intensification theory, but also about process innovation. It is action-oriented, from idea to commercial-scale implementation. Or, in other words, it is not just descriptive, but prescriptive.

## 1.2 Setup of this book

This book has the following setup:

Section A describes the theory of process intensification design and process innovation practices. Chapter 2 describes process intensification design theory. It presents an introduction to process intensification and its different design domains. Chapter 3 discusses the importance of each sustainable development goal for process innovation. Chapter 4 offers the VIB method: the three different perspectives in understanding industrial innovation practices – "V" stands for values of engineers, "I" for interests of stakeholders, and "B" for beliefs in society. Chapter 5 offers a description for multidisciplinary assessments for stage-gates.

Section B presents an application of the provided theory for the stage-gate innovation trajectory. It offers guidelines and methods for every stage. Chapter 6:

https://doi.org/10.1515/9783110657357-001

the discovery stage, Chapter 7: the concept stage, Chapter 8: feasibility stage, Chapter 9: the development stage, Chapter 10: the engineering procurement and construction stage, and Chapter 11: the implementation stage.

Section C offers five descriptions of industrial process intensification cases. These cases are structured in agreement with the theories presented in Part A and B. Chapter 12 presents a summary of the industrial case studies. All contain complete descriptions from idea generation to final commercial-scale implementation. Chapter 13 describes the BTG-BTL biomass to liquid fuels case. It is about modular design and prefab construction. Chapter 14 is devoted to the bulk chemicals Eastman methyl acetate multifunctional integration case with explicit function integration design and modeling aspects. Chapter 15 is about the Shell OMEGA Only Monoethylene Glycol Advanced process, applying a plant-wide process-intensified design approach. Chapter 16 is about intensified ethanol production by cyclic distillation. Chapter 17 presents a process-intensified pharmaceuticals case.

Part D is about education of engineers, both academic and industries. Chapter 18 describes BSc education in process intensification design. Chapter 19 describes industrial education of process-intensified innovation.

## 1.3 Wording

This book uses several general wordings to keep the style simple and easy to read. For instance, it uses the word "engineers" for all employees working in R&D departments in industries. In nearly all cases, it uses "he" where "she" could also have been stated. It uses one type of name for the innovation stages, although different industry branches use different names for stages and small companies often have no formal innovation stages.

## Reference

[1]  United Nations, Transforming our world: The 2030 agenda for sustainable development, 2015.

Part A: **Theory**

# 2 Process intensification design

**Abstract:** This chapter focuses on process-intensified design methods. Furthermore, it clarifies the differences between conventional process design based on established predefined unit operations and design based on process-intensified principles.

Process intensification (PI) design methods are not only suited for new processes but also for retrofitting and revamping existing processes. Special attention is paid to this subject and opportunities to apply PI are shown.

PI is a set of radically innovative process-design principles which can bring significant benefits in terms of efficiency, cost, product quality, safety, and health over conventional process designs, when compared with designs based on established unit operations.

All PI design methods can be classified into five domains: spatial plant-wide, functional synergy, temporal, thermodynamic, and spatial equipment. For each domain, guiding design principles as also a selection of established process-intensified technologies are described. The selection is limited to technologies for which all necessary concept design information is available.

## 2.1 Overview of process intensification design

### 2.1.1 Definition of process intensification design

Process intensification (PI) is a totally new way of designing processes. It can be considered as a radical change in the history of process design, similar to the radical change induced by the introduction of unit operations by Arthur D. Little in 1916, to process design. That introduction was then seen as the beginning of chemical engineering as a specific discipline [1].

Prior to the introduction of unit operations, process designs were specific for the materials to be processed, so there were sugar process technology, coal process technology, oil refinery technology, and so on. After the introduction of unit operations, generic knowledge and design methods were developed for each unit, placed in Perry's Chemical Engineer's Handbook [2] and so became a common practice for process design, regardless of the specific industry branch. That handbook was regularly updated to include new operations and improved design methods. In the eighties, computer program modules became available for most unit operations, further advancing process design based on unit operations.

A unit operation is a basic step in a process defined by physical, chemical, or biochemical changes and governed by a set of design methods which are applicable to all materials to be processed, for which the physical and chemical properties are known. Very complex and large process designs can be easily made by combining these unit operations. First, the whole process can be designed with unit operation

https://doi.org/10.1515/9783110657357-002

blocks connected by streams. For each unit operation, specialists can make the unit operation design, in detail. In this way, process designs can be made quickly and reliably. This process-design method has become the prevailing design method not only for chemical processes, but also for most other processes such as food and minerals processes.

PI design is breaking away from unit operations design. Keil discusses various PI definitions in his magnificent review article and concludes with the European Roadmap of PI definition for Process Intensification: "Process Intensification is a set of often radically innovative principles ("paradigm shift") in process and equipment design, which can bring significant (more than factor 2) benefits in terms of process and chain efficiency, capital and operating expenses, quality, wastes, and process safety" [3]. The same definition appears in the latest textbook on PI [4].

The remarkable thing is that this definition, as with all other PI definitions reported by Keil, does not specify upon which subject the benefits are made, or on what existing paradigm the shift is made. The most obvious existing paradigm from which PI shifts away, but is not explicitly stated is, of course, that of process-design based on unit operations.

With this insight, we refine the definition of PI as follows:

**Process intensification is a set of radically innovative process-design principles which can bring significant benefits in terms of efficiency, cost, product quality, safety and health over conventional process-designs based on unit operations.**

Reduction in size can also be a benefit of PI, for instance, when it concerns a retrofit of an existing process plant, or when otherwise large equipment has to be transported. But size reduction is, in most cases, a minor benefit and is therefore not part of the definition.

Stankiewicz [4] also provides the following principles for PI:
1) Maximize the effectiveness of intra and intermolecular events;
2) Give each molecule the same processing experience;
3) Optimize the driving forces and resistances at every scale and maximize the specific surface areas to which these forces or resistances apply;
4) Maximize the synergistic effects from partial processes.

These principles are very abstract. Here, they are clarified and also indicated how they are treated in this book.

Intermolecular events of principle 1 are about the interaction between molecules; hence, the selection of chemical reaction system for the desired product. This subject is treated in detail in section 2.2.1. Intramolecular events concern phenomena inside the molecules, such as the attenuation by radiation, as in a photochemical reaction. This subject is treated in Section 2.5: Thermodynamic domain.

Principle 2 concerns the residence time distribution (RTD) inside process equipment. The principle states that plug-flow is desired. However, this is not always the best potential solution for a given process. Some reaction systems such as autocatalytic reactions benefit from some degree of back-mixing. The subject of RTD is treated in every section of this chapter, where it is relevant.

Principle 3 concerns the optimization of heat and mass transfer inside process equipment. Heat transfer is treated in some detail, in Section 2.5 and also in other sections where relevant. Mass transfer is treated in Section 2.6 on intensified process equipment.

Principle 4 concerns the synergy between sub-processes inside a process unit. This is treated in detail in Section 2.3 about finding synergy effects by combining functions.

PI design methods can be applied to any process-design problem in any process industry branch. In the end, processes are artifacts that transform feedstocks to desired products. Typical transformation functions are mass movement, heat exchange, mass exchange, mixing, dispersion, reaction, separation, grinding, and shaping [5]. For all these transformation functions, PI design methods are available, and in several cases, process-intensified equipment is also available. Classification of PI methods and equipment is treated in the next section.

## 2.1.2 Classifications of PI

Keil [3] provides an overview of several PI classifications. From the overview, it is clear that the most useful classification for design approaches is domain classification of Stankievicz in his textbook [4]. The PI domains are: spatial, functional synergy, temporal, and thermodynamic.

The spatial domain has two subdomains; the plant-wide spatial domain and the equipment spatial domain. As these two subdomains are very different in design scope, we will treat them as different domains. For process-design purposes, the most logical sequence is then to start with the largest design scope namely plant-wide, and end with the most narrow design scope, equipment. Function synergy is nearest to plant-wide, while temporal and thermodynamic are similar in scope. The most logical sequence of the five PI domains for design is as follows:
1)  Spatial plant-wide
2)  Functional synergy
3)  Temporal
4)  Thermodynamic
5)  Spatial equipment

The remainder of this chapter is structured along these five domains.

### 2.1.3 Opportunities PI applications

#### 2.1.3.1 Radically new processes

PI can be applied whenever a new breakthrough process is desired. Then all PI principles and technologies can be considered to arrive at new concepts which are substantially better than a conventional process-design based on established unit operations.

#### 2.1.3.2 Modular design processes

Small-scale distributed processes gain more and more place in process industries, particularly for biomass conversions and for manufacturing precursors at the client premises. The main motivation for these plants is to avoid costly transport (wet bulky biomass) or to avoid transport of dangerous precursors (such as chlorine).

Small-scale plants have the disadvantage of high investment cost per ton of product, which can be calculated from the cost-capacity 0.6 power law curve. However, two methods are available to counteract these higher costs. The first method is modular design. In this method, the process is designed as a module that is copied several times so that design cost occurs only once for many identical plants distributed across many locations. The second method is prefabrication of the process at the construction firm site. It is tested there and then transported to the production site. The prefabrication at the construction firm site means that construction is executed by well-trained engineers at a controlled environment, with all tools directly at hand. Moreover, precommissioning tests (nitrogen flush, water flush) for equipment and control integrity are carried out at the construction site and errors are quickly rectified.

Chapter 16 shows a biomass case of this modular design and prefab construction and implementation.

#### 2.1.3.3 Retrofitting (revamp) opportunities and methods

PI can also be considered to revamp, retrofit, or debottleneck the existing processes. In my career, I have been involved in many revamps and debottlenecking of existing processes. These always have very high returns of investment (short payback times). In some cases, when revamp seems to be impossible because of the lack of plot space, PI may be the only way to find a solution. Here are a few examples followed by a general procedure for revamping.

The first RD applications of CDTECH were revamps inside oil refineries to obtain a higher quality transport fuel. The revamp solution was placing bags containing catalyst particles on distillation trays and feeding hydrogen to the column, so that hydrogenation reactions to the desired product occurred; later, by distillation, the upgraded hydrocarbons to transport fuels were obtained [5].

Revamps of reaction sections are particularly attractive, as often, less by-product is made, so that separation sections have a lower feed and debottlenecking of the whole process can be obtained. In general, these types of revamps have a high return of investment.

Batch processes can often be de-bottlenecked by a PI technology in combination with a change to continuous operation, such as the liquid-liquid reactive extraction case described by Harmsen [6].

Distillations can be debottlenecked by applying a dividing wall column. Due to the increased efficiency, vapor flows and reboiler duties are reduced, creating room for debottlenecking [7].

Distillations can also often be debottlenecked by connecting a membrane. The membrane can reduce the reflux required, by which the vapor flow in the column is reduced and the bottom boiler has a lower duty. This, then creates room for debottlenecking.

PI technologies can also be installed as an add-on. If, for instance, due to more stringent environmental regulations, emissions have to be reduced, a process-intensified technology as an add-on should be considered. PI technologies have a low foot print, so often, they can be placed in available small process spaces. A classic example is the reverse flow combustor to remove volatile organic components (VOC) from flue gas so that more stringent environmental regulations are met.

It is not easy to provide general guidelines for revamping existing processes. Each process has its own context and opportunities. Also, the need for revamping can be very different. In some cases, it will be the need for a higher production capacity. In other cases, it will be the need to comply to more stringent environmental regulations. In yet other cases, it may be the need for a more pure or modified product.

It is advised to formulate the revamp goal and then follow innovation stages from discovery to start-up, so treat it as a new process innovation [8]. Further information on revamps and retrofitting is provided by Rangaiah [9].

## 2.2 Spatial domain plant-wide PI

### 2.2.1 Plant-wide intensification principles

The notion of plant-wide intensification is obtained from Portha [9], who distinguishes global (plant-wide) from local (equipment) PI. Plant-wide PI is then about reducing the size of the whole process space and to make it more efficient.

The main guiding principle for process-wide intensification is: Avoid by-products in reactions. By this principle, the number of separations are reduced and thereby, the process as a whole is intensified, that is, the overall plant size is reduced, investment costs are reduced, and primary energy needed will be reduced. The latter is due to the fact that every process step is irreversible: high quality energy (exergy) is needed to drive the process step. Hence, reducing the number of process steps always goes hand-in-hand with primary energy need reduction.

By-products are made when the reaction stoichiometry inevitably produces by-products or when the reactions are not selective. Chapter 15 shows an example. The first is also the case when auxiliary reacting agents are applied. These are also called stoichiometric reagents [10]. These reagents are often applied in fine chemicals and pharmaceuticals to obtain the selectively desired product. However, the auxiliary reagent needs to be removed in a subsequent step, needing an additional reaction and an additional separation step while creating a waste stream. Avoiding these reacting agents is not easy. A catalyst will have to be found to facilitate the desired selective reactions. Books on green chemistry, such as by Anastas [10, 11], are a good entry to find catalytic routes rather than stoichiometric reagent routes.

The second case is when the chosen reaction is not selective and undesired by-products are made. Often this occurs when the reaction is a so-called thermal reaction, with no catalyst involved. A catalytic route should then be searched for, or an auxiliary reacting agent should be applied that can be easily recycled in the process. For instance, if the reagent is a low boiling component such as carbon dioxide, then decoupling and recycling can be done at low cost. Chapter 15 describes in detail the Shell OMEGA process case where catalysts and an auxiliary reagent (carbon dioxide) are applied, and both are recycled. By this combination, the reaction selectivity increased from 90% to 99.5%, by which a number of separation steps could be avoided.

There are various ways of changing the molecular interactions such that no by-products are made. Here is a list to consider in order to obtain the desired selective reaction.
– Apply catalysts
– Apply biocatalysts (enzymes)
– Apply microbiology
– Apply a recyclable protecting agent so that the undesired reaction does not take place: an inhibitor for unwanted reactions.

### 2.2.1.1 Catalytic reactions

Often, a heterogeneous catalyst that can be used in a fixed bed reactor is preferred. In this way, no catalyst recovery step is needed. However, highly selective homogeneous catalysts are preferred over less selective heterogeneous catalysts, because this means a lower feedstock cost and also less separations of by-products.

If a homogeneous catalyst cannot be avoided, develop a process concept in which the catalyst can be easily separated from the product stream. This may be obtained by using a solvent for the homogeneous catalyst in which the catalyst remains, while the product is separated from the solvent plus catalyst. This can be obtained by applying a solvent with a higher boiling point than the product so that the solvent plus catalyst can be separated by distillation.

### 2.2.1.2 Enzymatic reactions

Enzymes are catalysts from biological systems. They can be enormously selective for a certain reaction. To avoid a separation step to recover the enzyme from the product stream, the enzyme can be fixed on solid carrier material so that a fixed bed reactor can be obtained – with no significant loss of activity, in most cases. Enzymatic reactions are part of biotechnology conversions. Text books are available to find existing enzymatic reactions. Enzymes can also be modified to tailor them to a specific new reaction, such as a new route to nylon. The modification can be done by chemical means, or be part of microorganism selection or microorganism genetic modification. Frances Arnold is the founder of a totally new method of making enzymes to catalyze industrially important reactions, not known in nature. Here, the method consists of modifying the DNA structure of microorganisms a little bit, so that the microorganisms make modified enzymes. Chemicals are then fed to the enzymes and the chemicals produced are screened.

Nowadays, making small DNA modifications in microorganisms is very easy because kits that do this are available. It takes only a few days to get the modified microorganism and perform the screening for chemical activity. As the enzymes are highly selective, including enantioselective, new chemicals can be made in a single step at high selectivity. Good introductions to the subject are provided by Hammer [12] and Arnold [13]. A powerful technique to exploit this is high-throughput screening.

### 2.2.1.3 Microbiology reactions

Microorganisms carry out a chain of reactions inside their body. The overall reaction is often very selective and recovery of the organisms from the product stream can be done by filtration. Classic cases are ethanol from sugar and methane from organic waste. However, there are text books filled with microbial conversion steps, such as the one by Lee [14].

### 2.2.2 Application areas for process wide PI

The obvious application area for process-wide PI is for processes with many steps. Reducing the number of steps is then the objective. The advantages of reducing the number of process are lower investment cost, lower operating cost, lower energy requirements, lower fugitive emissions to the environment, and ease in meeting safety criteria.

Lower investment cost is due to fewer major process equipment, as also less piping. In general, piping accounts for 20% of the investment cost (see *Perry's Chemical Engineers' Handbook*) Hence, reducing piping always means lower investment cost.

Reducing the number of process steps also means lower maintenance cost as well as lower other operating costs.

Each process step is thermodynamically irreversible, hence requires high-quality energy. Reducing the number of process steps will reduce the amount of high-quality energy, nearly always.

Fugitive emissions are caused by leaking flanges to connect equipment. Fewer equipments to connect means to lower diffusive emissions.

A lower number of process units also means a safer process, as a process step that is not there cannot be unsafe. This is one of the methods for an inherently safe design [8, p. 260]. It also means fewer flanges, which, in turn, means fewer of potential leakage points.

## 2.3 Function integration domain

### 2.3.1 Introduction of function integration

Process functions describe what needs to be done, not how they are done. Thinking in terms of functions rather than in terms of unit operations is probably the largest shift in designing processes in the last fifty years. The godfather of this shift is Jeoffrey Siirola, who introduced functions for concept design (rather than unit operations) in the 1970s [15, 16].

The most common functions in processes are as follows:
–  Mass movement
–  Mixing
–  Mass transfer
–  Enthalpy exchange
–  Reaction
–  Separation
–  Shaping
–  Size reduction

By these functions, feed streams can be transformed to product streams. A function design method is provided in Section 2.3.2.

The advantages of designing with functions are two-fold. The first is that the functional design creates many opportunities to integrate functions. The second advantage is that for each function, the best technique can be chosen.

### 2.3.2 Guiding principles function integration

#### 2.3.2.1 Opportunities

In general, opportunities for obtaining function integrated designs are many. In particular, when the process contains mainly gas and liquid streams, several options can be generated. But when solids are also involved, function integration options can be generated as shown in the BTG-BTL case of Chapter 13. In fact, designing with functions should always be done in combination with plant-wide design.

The advantages of designing in the functional domain with functions and streams are:
– Only essential functions are defined, functions can be combined in a single apparatus.
– Several equipment choices can be considered to fulfill the function and the best selected.
– The number of process steps is reduced.
– The size of the process steps is reduced.
– The exergy loss is less due to fewer steps and due to a more even driving force distribution inside the process step [17].

#### 2.3.2.2 Concept design method

Process concept design starts with defining the product type and its composition and the feed types and their composition. Then, the differences between feed and product are eliminated by identifying and choosing functions and connecting the functions by streams so that in the end the feeds are connected to the product by streams and functions.

If the product has molecules that are different from the feedstock then of course a chemical reaction is needed to go from the feedstock to the product. For the time being, no further detail is required – just a function block reaction is placed in the design solution space. If a separation is needed, then a separation block is placed in the solution space. In this way, all essential functions are defined in the design solution space.

A particularly interesting aspect of function integration is the use of process streams. Siirola considers streams not just connections between block functions, but available means that can be used to perform certain actions. In his classical, methyl acetate process streams are used to extract and absorb components from other streams. In this way, Siirola reduced the conventional process design of 11 process steps to a single RD column [18]. A full description of that process design is found in chapter 14 of this book.

In general, process streams, including feed and product streams, can be used as a means for:
- Solvent
- Extraction
- Absorption
- Desorption (anti-solvent)
- Adsorption.

Then, options for the reaction function are further evaluated. Section 2.2.1 is helpful in generating reaction options. After that, separation options are generated. Textbooks on process concept design such as those by Douglas [19], Seider [20], and Harmsen [8] will be helpful in this respect.

After that, combinations of reactions and separations are explored to see whether a window of pressure and temperature can be created for both reaction and separation. If so, then these functions can be integrated into a single piece of equipment. For RD, Section 2.3.4 provides further design guidelines.

Stankiewicz provides many examples of function integration [4] which will be helpful in further exploring options. Also, the case descriptions of BTG-BTL (Chapter 13) and Eastman (Chapter 14) may be helpful in generating options.

Here is a list of mostly applied function integrations in industry, provided by Qi [21]: reactive distillation (RD), dividing wall column (DWC) distillation, membrane assisted distillation, membrane assisted crystallization, pervaporation, reactive extraction, extractive distillation, reactive absorption, reactive chromatography, membrane reactor, reactive crystallization, reactive precipitation, reactive grinding, reactive filtration, and reactive extrusion.

In the next section, two types of function integrations are treated in some detail, because these options have been applied so frequently that they have become established unit operations.

## 2.3.3 Reactive distillation

Reactive distillation combines the functions of reaction and distillation in a single column. The reaction can be thermal, homogeneously catalyzed, or heterogeneously catalyzed. In the last case, the catalyst is fixed in the distillation column. Many process applications of this function-integrated design are described by Harmsen, notably for hydrogenations (of crude oil fractions), esterifications, and etherifications, and mostly commercially implemented in oil refineries and bulk chemical plants at very large scales: 100–3,000 kton/year [22].

Selection criteria for RD are [23, 24] as follows:
1) The reaction is so exothermic that cooling is needed and the reaction temperature is compatible with distillation. By RD, the reaction heat is directly used for

evaporation. The distillation temperature can often be changed to the reaction temperature by changing the distillation pressure. The main advantages are investment cost savings, operating cost savings and lower diffusive emissions due to less flanges. All RD applications for hydrogenations in oil refineries have been installed because of this rule.

2)   When the main reaction is an equilibrium reaction. By distillation, the product is removed from the reaction location and the reaction can be completed inside the distillation column. Typical examples are esterification and etherification reactions.

3)   When the distillation has an azeotrope, or is a close-boiling system. By reacting away one component the other component can be distilled off.

4)   When solvent extraction or stripping has a beneficial effect on the distillation. In this way azeotropic distillation can be avoided.
     The Eastman Chemical case described in Chapter 14 shows the large benefits of applying rules 2, 3, and 4.

5)   When the catalyst is fouled by a heavy boiling component. By feeding below the catalyst section, the heavy boiler does not reach the catalyst and longer catalyst life is obtained [22].

Concept design methods are provided in text books. The most recent one is by Luyben [25]. Detailed design knowledge is available from the technology providers CDTECH and Sulzer Chemtech.

Scale-up methods for reactive distillation are brute-force scaled-down pilot plant and model-based pilot plant design [6]. The brute-force method involves a scaled-down pilot plant with all conditions and parameters being the same as in the commercial-scale design, except for the column diameter. This method is mostly used in oil refinery applications where detailed kinetic information of all reactions is not available. The model-based scale-up method involves a scaled-down pilot plant where the structured catalytic packing is reduced in size so that the pilot plant height is also reduced. The mass transfer and RTD characteristics of the packing are known, so the design process model can be falsified or validated with pilot plant test results.

### 2.3.3.1 History of RD implementations

RD has been known in the base chemicals industry for over 60 years. The first commercial-scale application inside Shell was in 1953 [26]. Eastman Chemical installed a large reactive extractive distillation for the production of methyl acetate in 1983 and also explained the complete scale-up method and history [27]. These reactive distillations however, did not receive any significant follow-up in the petrochemicals and oil refinery industry.

This changed when the technology provider, CDTECH, brought their catalytic distillation technology to the market. Their first commercial-scale implementation was at the Shell Motiva oil refinery in 1987. It was a revamp of an existing distillation column, using the Texas tea bags catalyst technology of CDTECH. Heterogeneous catalyst particles inside porous bags were placed on the distillation trays. The reaction involved was a hydrogenation [22].

By 2007, the technology provider CDTECH had a total of 123 implementations, of which 60 were etherifications [22]. On their present website, they report more than 100 implemented etherifications [28]; an increase by 40. For the other applications, the CDTECH website provides no new data. This means then that CDTECH has installed more than 160 reactive distillations.

The technology provider Sulzer does not provide numbers of implementations on their website or otherwise. They have provided me with details of implemented RD per process type and per application the region of implementation, in 2007 [22]. If each process type and region is installed at least once, they have installed at least 10 commercial-scale RDs. It is interesting to note that BASF has installed more than 50 RD [29]. Whether this is all by in-house development or whether some have been obtained from technology provider Sulzer is not reported. It can then be concluded that the total number of RDs installed by the CDTECH and Sulzer Chemtech is well over 170. This finding indicates that for oil refining and petrochemicals, having a technology provider increases the rate of commercial-scale implementations enormously.

### 2.3.4 Dividing wall column distillation

The DWC distillation combines two thermally coupled distillations in a single column. The column has a partly dividing wall inside such that either the top, or bottom section, or both of the two distillations are common [30], so it is applied when at least three product streams are to be produced by distillation.

A selection criterion for considering a DWC is when the middle boiler is the largest of the three product streams according to Schoenmakers (e.g., BASF) as reported by van den Berg [31].

Concept designs are best made by first treating the process design of two distillation columns and then combining the columns into one with a shared top section and/or a shared bottom section [31].

#### 2.3.4.1 Applications dividing wall column distillation

The first commercial-scale implementation was in 1985 by BASF [32]. Since then the number of commercial implementations has grown strongly, in the last few decades. The technology provider Montz reported over 90 implementations [33] in 2010, and Sulzer Chemtech reported in 2012, that it had 36 DWC implemented. [34]. Schultz had reported already in 2002 that some DCW implementations were by engineering firms other than Montz or Sulzer [32]. The present total number of implementations is therefore, well in excess of 130.

The driving forces for implementation are energy savings and investment cost savings [28]. The energy savings typically range from 10% to 30% [31]. The investment savings are typically 30% but in specific cases can be 60% [32].

The basic design and control knowledge is available in the public domain [35, 36]. Detailed design for the construction is available from technology providers Montz [33] and Sulzer [34]. An overview is provided by Dejanović [36].

The scale-up method applied is model-based. The model is validated in a mini plant or a pilot plant [31, 33]. The largest commercial-scale DWC is operated by Sasol, with a height of 64.5 m and a diameter of 4 m [31].

## 2.4 Temporal domain

### 2.4.1 Guiding principles temporal domain

The temporal domain is about applying dynamics on purpose to obtain an efficient process. In the temporal domain, the concept designer is focused on using variations in time, such as cyclic operations, pulsed operations, and reverse flow operations.

The main guiding principle for applying the temporal domain is decoupling phenomena such that each phenomenon can be optimally designed for the desired process task. Hence, if several phenomena are needed for a process step and are coupled in a conventional process unit which is very expensive, then designing for the temporal domain should be considered.

Stankiewicz provides many technologies for this temporal domain [4]. We have selected three technologies to illustrate the application power of this domain. Two are described in the following sections. Cyclic distillation is described in Chapter 16.

### 2.4.2 Reverse flow reactors

A reverse flow reactor is a fixed bed adiabatic reactor in which the feed location (and also therefore, the outlet location) is periodically changed from one end to the other end. In this way, the middle part of the reactor can operate at a very high temperature (typically 500 °C), while the outlet temperature is only the inlet temperature plus the adiabatic temperature rise (typically 40 °C). Hence, a reverse flow reactor is a process-intensification technique operating in the temporal regime. It combines the functions of heat exchange and reaction in a single vessel. The periodical operation causes an effective counter current heat exchange effect, without having to install heat exchangers at the front and back of the reactor. In this way, investment costs are kept very low compared to a conventional set-up of a reactor with a heat exchanger that recovers the heat from the outlet and uses it to heat the feed stream.

The first implementation was in 1982 [37]. In 1993, over 10 processes were in commercial operation [38]. In 2012, thousands of these reverse flow reactors were in commercial-scale operation [39].

Reverse flow reactors are applied for dilute gas systems (with a limited adiabatic temperature rise) [38]. Applications are $SO_2$ oxidation, $NO_x$ reduction, $N_2O$ decomposition, and VOC oxidation [37]. Either heterogeneous catalyst or inert ceramic material is employed. For VOC oxidation, Bunimovich shows the advantages in cost when using a catalytic system [40]. For a VOC oxidation of a gaseous effluent of an ethylene oxide plant, McNamara describes a system with ceramic balls [41]. There are several technology providers for this well-established technology [37].

The driving forces for these applications are environmental legislation to remove contaminants in effluent gases to the atmosphere and to reduce investment cost, compared to conventional solutions with a heat exchanger [38]. The investment cost is typically reduced by a factor of 4, and in some cases, energy is also saved [38].

Design and modelling knowledge of this unconventional technology is, of course, key to commercial-scale implementation. This knowledge is available at Matros Technologies [42]. A very good description of the design principles for reverse-flow reactors is provided by Eigenberger [43]. Specifically, for hydrocarbon combustion, Marin provides a simplified "pseudo steady state" design solution [44].

> The main drawback of the reverse-flow reactor in comparison with steady-state conventional designs is that the conversion losses during the flow reversal procedure resulting from a possible reaction mixture bypassing the reactor during valves switching, and with replacement of unreacted mixture from pipelines reactor void volume and inlet heat regenerative bed (before the catalyst bed) into the outlet stream immediately after flow switching. Losses from valve leakage may be minimized by application of quality valves with low duration of the switching, while the losses from mixture replacement may require the changes in the reactor design (minimization of void volumes) and the improvements in the process flow sheet, e.g., by application of various purging circuits and intermediate vessels for gas storage during switching under appropriate control strategy [45].

Another problem related to VOC combustion is condensation of water that disrupts the heat balance. By adding adsorption material at both sides of the packed bed, this problem can be prevented [46].

Scale-up has been carried out by modeling and validation by pilot plant tests [38, 47].

### 2.4.3 Oscillatory baffled reactors

Oscillatory baffled reactors are tube reactors with ring baffles at the wall and have an oscillatory back-and-forth flow super-imposed on the net flow through the tube.

The oscillatory flow, in combination with the baffles, creates small-scale turbulence, enhancing mass transfer and heat transfer at a net low averaged velocity, through the tube reactor. The technology is well-summarized by van Gerven [4].

This technology is attractive for reactions and crystallizations requiring a long residence time (more than an hour), turbulence (for 2-phase mass transfer and heat transfer), and a near plug flow RTD.

Often, these reactions and separations are performed in mechanically-stirred reactors that are batch-operated. The oscillatory baffled reactor will result in lower investment and lower maintenance for crystallizations producing more uniform particles.

Concept design knowledge for sizing this technology for gas-liquid, liquid-liquid and liquid-solids systems is available from McDonough [48].

The scale-up method can be brute-force numbering up of tubes in parallel. However, then uniform distribution over the tubes is a challenge. It can also be done by validated dimensionless number correlations, as recommended by McDonough [48].

Here is a potential application case for emulsion polymerization to show the power of the technology. Emulsion polymerization requires homogeneous turbulence for uniform droplet dispersion, plug-flow for deep conversion, and high heat transfer for cooling. The conventional design is a mechanically-stirred batch reactor with cooling by evaporation of a solvent. This design is very expensive and reliable scale-up ensuring the same product quality is very difficult than in general empirical scale-up with several intermediate-scale pilot plants.

The temporal domain solution is a baffled pipe reactor with oscillating flow that operates continuously. The oscillating flow causes nearly uniform turbulence for uniform droplet formation, high heat transfer, and plug flow. The averaged linear velocity and the pipe reactor length can be chosen such that a multitube reactor is obtained. A down-scaled single tube pilot plant is then sufficient for a reliable scale-up design and implementation.

## 2.5 Thermodynamic domain

### 2.5.1 Thermodynamic domain guiding principles

Thermodynamics is about energy and transfer. In thermodynamics, energy has two aspects: the amount of energy and the quality of energy. High quality energy, called exergy, is needed to drive processes. While this is happening, the quality of the energy supplied is lost. Forms of exergy are: Work, electricity, radiation, high temperature heat and chemical free enthalpy. As every process step is irreversible, it always means loss of exergy. However, the degree of irreversibility can be influenced by design and so can the amount of exergy loss.

The form in which high quality energy is to be supplied to the process step at hand can be chosen. In conventional process designs, high quality energy required to drive reactions is supplied by the negative free enthalpy of the feed components. The same holds for mass transfer processes. For endothermic reactions, high temperature mediums such as hot gas or steam are applied. For mechanical process steps such as pumping, high quality energy is supplied by electricity.

In the thermodynamic domain, a whole array of high quality energy types to drive processes are provided by the textbook of van Gerven [4]. Inspired by this textbook and the industrial applications, I derived the following three guiding principles.

Guiding principle 1: Consider supply of high quality energy by light radiation, microwave radiation, sonic waves, or electricity if the medium requiring the energy is a solid or a highly viscous liquid. The latter can be applied to charge particles (electrostatic particles separations), to heat (by induction), or by electrochemistry.

Guiding principle 2: Consider the thermodynamic efficiency of the whole system of primary energy input, the process step, and the outputs of the system to find a PI solution.

A beautiful industrial-scale case of this guiding principle is the BTG-BTL biomass pyrolysis process described in Chapter 13. In this process, rapid heating up is obtained by hot fluid bed particles. The hot particles are obtained from fluid bed combustion of char.

Guiding principle 3: Consider new ways of removing enthalpy from the system (cooling).

Process intensified cooling can then be done by flash evaporation or fluid bed (solids) cooling or by high heat transfer equipment (micro tubes). Rapid cooling can be important to preserve product quality (sub-sequent reactions are then avoided).

Due to the driving force to find renewable energy sources, rather than fossil sources of the process industries, many R&D programs are working on exploring alternative energy sources and alternative ways of supplying energy to the processes. However, so far, there have been only a few commercial-scale applications of technologies for the thermal domain. The main application is electromagnetic heating. This is described in the next section.

### 2.5.2 Electromagnetic (microwave) heating

Electromagnetic heating is based on electromagnetic radiation attenuating molecules with a dipole moment (permanent or induced). In this way, high heat transfer and more uniform heating is obtained, compared to heating based on conduction. This is very attractive, especially for solids and highly viscous liquid such as food materials, where rapid heating is beneficial for food quality and is applied in many food factories.

The theory of electromagnetic heating is well established and the dielectric properties of most material are known. Small-scale experiments can be done to validate the heating up performance. Scale-up can be done by keeping the treatment layer the same as scale-up from the pilot plant. Continuous operation is often applied by using a moving belt on which the material is placed. A recent textbook on food processing with all needed design and selection information as also many industrial-scale applications is available [49].

## 2.6 Spatial domain equipment

Spatial domain equipment design is about reducing the equipment size as compared to conventional equipment. For selection and design purposes, two steps are described here. The first step is defining the criteria required for the process equipment to be selected. Criteria guidelines for this step are found in Section 2.6.1. The second step is to choose equipment with these guidelines. Section 2.6.2 provides many process-intensified equipment options with attributes in the same terms as the selection criteria, so that selection is facilitated.

### 2.6.1 Spatial domain equipment intensification criteria

Harmsen provides a list of critical performance phenomena for processes [6]. These are used here as a starting point for defining PI criteria guidelines.

Guideline 1: **Optimize residence time distribution**
Select the best RTD for the given reaction system. This can be plug-flow, back-mixed, distributed feed, or a combination of these. Most reaction systems benefit from plug-flow, in which the required average residence time to obtain deep conversion is short. This, in turn, also minimizes consecutive by-product formation. Some reaction systems, however, benefit from back-mixing behavior. These are reaction systems with "autocatalytic behavior". A classic example is microbiological growth systems, in which the cell growth rate is proportional to the cell concentration. In a back-mixed reactor, the cell concentration is the same as the outlet concentration. In a plug-flow reactor, the cell concentration is much lower than the outlet concentration everywhere. The latter reactor thus requires a much larger volume than the back-mixed reactor for the same outlet concentration and production rate (kg/h).

Another example is peroxide formation of an intermediate which catalyzes the feed component to form this peroxide. Textbooks on reaction engineering provide methods to define the optimum for a given reaction system, such as by Levenspiel or Scott-Fogler. For separations and grinding, nearly always, plug-flow is the optimum criterion.

Guideline 2: **Maximize mixing**

Some reaction systems are very sensitive to the rate mixing of the reaction components. They can also be sensitive to the rate of mixing with the bulk fluid. If the intrinsic reaction times to achieve some conversion are less than a few seconds for gas phase reactions and less than 1 min for liquid phase reactions, then micro-mixing effects should be further investigated with a simple model [15]. If that investigation reveals that micro-mixing rates are important, then reaction equipment with rapid micro-mixing attributes should be used. The textbook on PI by Stankiewicz is a good starting point [4].

Guideline 3: **Minimize mass transfer limitation**

For reactions avoid mass transfer limitations, particularly, when reaction occurs in the boundary layer. Otherwise, it would be dominant and would cause selectivity losses.

By this principle, equipment size is reduced and also the exergy loss accompanying mass transfer is avoided. Avoiding mass transfer means that the process equipment becomes simpler to construct and smaller in size and requires less energy. This is because mass transfer requires a driving force, which means loss of free enthalpy (loss of energy), and requires the creation of interface area and mass transfer coefficient, which also require primary energy.

For two-phase separations, mass transfer limitations nearly always play a role. Mass transfer limitations can be reduced in several ways. For gas-liquid contact systems, it can be avoided by having a structured packing in a column or pipe, or by two-phase Taylor flow in milli-channels. It can also be established by avoiding the second phase, for instance, by using a solvent in which the reactant is dissolved. The solvent should preferably be obtained from available streams in the process concept design.

Guideline 4: **Minimize heat transfer limitations**

Often, in mechanically stirred batch reactors, the residence time is much longer than needed for the reaction, due to a heat transfer limitation. By removing this limitation, the residence time can be shortened, resulting often in by-product reduction. This, in turn, means lower feedstock cost per ton of product, lower separation cost, and lower energy cost.

Guideline 5: **Optimize for co-, cross-, or countercurrent flow**

For two-phase separations, countercurrent flow of phases is often very beneficial for deep removal of a component in a single equipment piece.

For reactors, no simple guideline can be provided. In some cases, cross-flow is the best choice of contacting two flows. Everything depends on the reaction system involved.

Guideline 6: **Uniform shear rate distribution**
Uniform shear rate distribution will cause a more uniform dispersion for two-phase systems, and thereby, a more effective mass transfer. In grinders, it will help in a more effective grinding. This also links strongly to two-phase reactor design.

Guideline 7: **Consider impulse transfer consequences**
Impulse transfer by the fluids on the equipment can be so high that internals and catalyst particles are damaged. This principle is often overlooked in equipment selection. Pressure drop information should be available for equipment to be selected.

### 2.6.2  PI equipment and attributes

Stankiewicz provides many equipment technologies [4] in his textbook. Here, a small selection is treated, that can be called established, as information on critical phenomena for selection and design information. For these too, scale-up methods are available and some have been applied in commercial scale. Table 2.1 contains these established equipment types with their key attributes that can be used for this selection. Scale-up information as well as industrial-scale applications are provided, where available.

**Table 2.1:** Established process intensification equipment types and their attributes.

| PI equipment | Phases | Mass transfer? and flow | RTD | Heat exchange | Mixing |
|---|---|---|---|---|---|
| Spinning disk | G/L, L/L, L/S | Yes Counter-current | Several stages | Yes | Yes |
| Rotating packed bed | G/L, L/L | Yes Counter-current | Narrow | No | Yes |
| Static mixer | G, L, G/L, L/L, L/S | Yes Co-current | Narrow | Yes | Yes |
| Monolithic structure | G, L, G/L | Yes Co-current | Narrow | No | No |
| Micro channel devices | G, L, G/L, L/L | Yes Co-current | Narrow | Yes | Yes |
| Impinging streams | G, L | No | Wide | No | Yes |
| Plate heat exchanger | G, L | Yes | Narrow | Yes | No |
| Membrane | G, L | Yes | Narrow | No | No |

Attributes looked for are:
- Phases that can be applied: gas, liquid, gas-liquid, liquid-liquid, solids.
- Flow modes: countercurrent, cross-current, co-current.
- RTD: narrow, some stages, back-mixed, wide.
- Heat exchange: present or not.
- Mixing: present or not.

All these equipment types can be applied as reactors. An example of applying these criteria for rotating reactors is provided by Visscher [50].

Most of them can also be used in separations. The descriptions that follow provide some more information for the designer to make a selection of equipment type for his design problem.

### 2.6.2.1 Spinning disk contactors

Spinning disks create a high centrifugal gravity acceleration field. This, in turn, also creates a counter current pressure gradient. This high gravity acceleration field, in combination with specific structural elements, creates high mass transfer and where applicable, countercurrent phase flow and high heat transfer. The high gravity field also allows for phase separations with small density differences. Here are three major spinning disk types already applied in commercial scale, with differences in flow and heat transfer attributes.

### 2.6.2.2 Podbielniak liquid-liquid contactor

The Podbielniak contactor is a liquid-liquid centrifugal contactor, that provides countercurrent flow with up to five theoretical stages in one unit and can separate liquids with a specific gravity difference as low as 10 kg/m$^3$ [51]. Scale-up information is most likely available from technology providers. It has been used for the first time in 1945 for penicillin extraction and since then successfully in the fine chemicals and pharmaceutical industries [51].

### 2.6.2.3 Thin film spinning disk contactor

The thin liquid film spinning disc reactor has countercurrent gas and liquid flow and heat transfer via the rotating disk wall to the medium on the other side of the disk material. The liquid residence time distribution is narrow. The gas phase RTD is unknown and could be wide. Ramshaw is the founding father of this technology [4].

The liquid side mass transfer coefficient is high, of the order of 0.0001 m/s, and the specific surface area is 1,000 to 100,000 m$^{-1}$, depending on the liquid viscosity. Hence, the overall mass transfer ($k_l a$) is up to 10 s$^{-1}$. The heat transfer coefficient is also very high; of the order of 20–50,000 W/m$^2$ K. Hence, the apparatus is very suitable for very fast gas-liquid reactions requiring high heat exchange rates [52].

The only commercial-scale applications reported are for pharmaceuticals production [53]. It has been established in one case that compared to a mechanically stirred batch process the reaction time is reduced by a factor 1,000 and the impurity by-product is reduced by 93%. A commercial-scale production of 8,000 kg/year was implemented. Scale-up using a reactor model appeared to be reliable. For polymerizations and precipitation reactions, lab-scale experiments are reported [54–56].

### 2.6.2.4 Rotor-stator spinning disk

The rotor-stator spinning disk (RSSD) contactor is different from the thin film spinning disk in that the gas phase flow is in a thin layer due to the presence of the stator plate, a small axial distance away from the spinning disk [57]. The RSSD is suitable for countercurrent gas/liquid and liquid/liquid flow. It has high mass transfer rates, rapid liquid phase micro-mixing [58] and high heat exchange rates via the spinning disk [57]. It also can be operated with fine particles in the liquid phase [58]. The liquid RTD is narrow [59, 98] and the liquid-liquid mass transfer rates are 25 times higher than packed columns [60]. The mass transfer ($k_{gl}a$) is typically a factor 100 higher than for other equipment. Moreover, the energy dissipation to achieve this high mass transfer rate is a factor 80 lower than for mechanically stirred tanks [20]. Also, the heat transfer rate is much higher than for other equipment [59]. Meeuwse provides an overview of this apparatus compared to mechanically stirred tank reactors for all these attributes and provides some examples of potential applications [59]. Liquid-liquid extraction has been patented [61] with flow profiles published [62].

Scale-up is mainly by numbering up as an increase in diameter means loss of the advantages on mass transfer [57]. This means that it is less attractive for bulk chemicals applications, as many devices have to be installed, then, creating a linear investment cost curve with capacity. For small capacity applications such as in fine chemicals and pharma applications, it is particularly attractive because it can handle viscous flows and has a high heat transfer rate. Hence, short reaction times are obtainable, that helps minimization of by-product formation.

### 2.6.2.5 Rotating packed bed

In a rotating packed bed (RPB), gas and liquid flow counter-currently at high g forces through a wire mesh packed bed, creating high velocities, high mass transfer coefficients, and high interfacial areas. Moreover, the phases show narrow RTD. Wang provides mass transfer information [63]. Wenzel provides liquid distribution and mixing information [64] and Yang provides computational fluid dynamic model and calculation results [65].

There are two obvious application fields for this technology. The first application field is for gas-liquid reaction systems where by-product formation is greatly reduced by a mass transfer coefficient ($k_l$). The higher investment costs are then easily counter-acted by the lower feedstock cost per ton of product. A typical application is hypochlorous acid (HOCl) production, where a high $k_l$ value means less by-product formation, as the consecutive reaction of the desired product with chlorine takes place in the liquid film. A high $k_l$ value means a lower concentration of the product in the film and thereby lower by-product formation. Dow has implemented a commercial-scale RPB process [66, 67].

A second application is in cases where the size of the gas-liquid contactor has a strong effect on the total investment cost of the infrastructure, such as gas or oil production platforms at sea. Applications in this field are mainly for gas purification in which the impurities to be removed are hydrogen sulfide, nitrogen oxide, carbon dioxide, sulfur dioxide, volatile organic compounds and nanoparticles [Guo], as also for de-aerating water [66].

Scale-up of a single RPB is limited, as with all rotating (spinning) contact equipment, due to the ever larger outer radius velocity at constant rotation angle velocity. Beyond a certain velocity, mechanical integrity of the equipment cannot be ensured anymore, so scale-up of a single RPB is limited, Often scale-up includes several RPB units. Dow mentions the following scale-up steps of a de-aeration case and the HOCl case [67].

Dow deaeration scale-up steps:
- Lab unit (scale not reported)
- Pilot plant: scale 50 t/h
- Commercial scale: 300 t/h
- Start-up information is not reported.

Dow hypochloric acid scale-up steps:
- Pilot unit 1/400 of commercial scale (single RPB)
- Commercial scale: 3 identical RPB
- Start-up is reported as easy

### 2.6.2.6 Static mixers
Static mixers are pipes filled with static mixer elements. These elements can also be heat transfer elements. The devices combine rapid mixing with plug-flow behavior and significant heat transfer rates. They are not only used in rapid liquid or gas mixing, [68] but are also used in co-current gas/liquid and liquid/liquid mass transfer, where plug flow is desired. The latter equipment is very attractive as a reactor for many applications [69, 70]. Additional advantages are that the mixers contain no rotating elements, which translates to high reliability, low investment cost, and low maintenance cost.

Micro-mixing rate information is available [68] and so are RTD, mass transfer and heat transfer [69, 70]. Suitable scale-up methods are brute-force and model-based.

In brute-force scale-up, all critical parameters – feed quality, mixer elements, number of feed points per cross-sectional area, feed flow per cross-sectional area (flow velocities), average residence time, temperature, and pressure of the pilot plant – are kept equal to the commercial-scale design. This scale-up method can be very attractive for pharmaceuticals and food applications, as it can be easily explained to the regulators that the commercial-scale design will produce the same product quality as the pilot plant.

In model-based method, the experimental results of the pilot plant are compared to model predictions. In this way, the design model is validated. The commercial-scale design is then carried out with the model. This scale-up method will be attractive for bulk chemicals as it keeps the pilot plant size and its cost small.

### 2.6.2.7 Monolith structures

Monolith structures consist of straight channels parallel to each other. They combine a medium-high porosity (70%) with a high surface area. The channels are very uniform in size because they are manufactured by extrusion. They are attractive for catalytic reactor applications that require low pressure drop and high outside-surface catalytic area. Often, fine catalyst particles are placed on to the structure using a so-called wash coating procedure to achieve this result. Monoliths are also attractive in co-current two-phase flow such as in Taylor flow applications [71].

Scale-up can be done by brute-force and model-based methods. For reliable two-phase Taylor flow scale-up, even flow distribution over the channels is important. How even distribution can be achieved is reported by Devatine [72]. Hydrodynamic and mass transfer information for model-based scale-up is provided by Haase [73].

Monolithic structures are applied as catalytic converters in commercial scale, in process industries, in flue gas treatment, and in exhaust pipes of cars for the removal of NOx and CO [71]. They couple a high gas/solid mass transfer with plug flow and low pressure drop. The high mass transfer and plug flow ensure deep conversion. The low pressure drop is needed as the up-stream furnace or engine can only tolerate a low overpressure [74].

Industrial-scale applications for gas-liquid flow in monoliths are still rare [75]. Manfe mentions a commercial-scale hydrogenation as part of a hydrogen peroxide process [74]. No other commercial-scale application can be found in the literature. According to Devatine, the reason for this low number of commercial-scale implementations is the lack of knowledge on how to achieve an even flow distribution over the channels [74]. Since his publication in 2017 on this subject, the number of commercial-scale implementations may have increased.

### 2.6.2.8 Micro-channel devices

Micro-channel devices have narrow channels causing high heat transfer, mass transfer, mixing and narrow RTDs. Established correlations for RTD, mixing, and heat transfer in pipes are applicable. Scale-up can carried out by numbering up the channels. Even feed flow distribution over the channels needs special distributor design [76]. Knowledge on the design and on manufacturing is available [77–79]. The investment cost shows a linear relationship with capacity [80]. This makes their applications attractive for small capacities such as in pharmaceuticals and fine chemicals, but not for bulk chemicals and other large-capacity applications. For micro reactors with heat exchange, the capacity point at which they exceed the investment cost per ton of product of conventional reactors is 40 MW heat exchange. Below this capacity point, their investment cost per ton of product is lower than that of conventional heat exchange reactors [80].

The application area for pharmaceuticals and fine chemicals is, therefore, treated in some detail. The most applied technology in these industries is the batch-operated mechanically-stirred tank reactor with heat exchange for heating and cooling via the reactor wall. Often, additional cooling is obtained by evaporating a solvent and applying an overhead condenser. The first advantage of micro reactors (and milli reactors) is that continuous operation at a very small scale is feasible, due to excellent mixing and heat transfer rates and a narrow RTD of the flow. The second advantage is that scale-out is fast compared to the batch operation, that often goes in an empirical way in steps from lab-scale, small pilot scale, larger pilot scale, and finally to commercial scale. In the continuously operated reactor, the lab-scale often produces sufficient material for trial tests, and scale-up to commercial scale can be done by numbering up.

The main reason for micro (an milli) reactor applications finding their way in the pharmaceutical industry very slowly is that the manufacturing of the new drug is highly regulated. It means that the manufacturing procedure once regulated cannot be easily changed. This, in turn, means that in the very early stage, the continuously operated micro reactor has be chosen and the regulators have to be convinced that the larger scale design will produce exactly the same product quality. AstraZeneca, GlaxoSmithKline, and Foster Wheeler have founded a nonprofit company Britest Ltd. to develop continuously operated processes and the manufacturing and scale-up procedures acceptable for the regulators [81].

The potential of these small devices to be applied in the pharmaceuticals and fine chemicals industries is large, as in the laboratory, the new molecules can be quickly synthesized to produce sufficient material for all kinds of tests. Jensen and Reintjens (DSM) provide a lot of information on specific devices used in this so-called flow chemistry [78, 82]. The term flow chemistry means that the synthesis of the desired molecules in the laboratory is carried out as a continuous process. The advantages are that highly exothermic reactions can still be carried out with a very short residence time, often in seconds, rather than in hours. In some cases, the separations are also carried out continuously in micro-channel devices. Compared to batch processing, the advantages are huge. In batch processes, overhead cooling of the reactions often takes hours. This means that by-product formation is reduced and a solvent for evaporation purpose is avoided [83].

### 2.6.2.9 Milli-channel reactors

Milli-channel reactors are the bigger brothers of micro-channel reactors. They can be designed more easily and constructed with two systems separated by a heat-exchange wall. In one system, a reaction takes place, while in the other system a heat medium flows, or a different endothermic reaction requiring heat occurs. Reaction heat is exchanged via the channel walls to the other side where a heat exchange medium also flows in narrow channels. Due to the small channel sizes, both the specific heat transfer areas and the heat transfer coefficients are very high. Because of this feature, very exothermic and very fast reactions can take place under nearly isothermal conditions. The reaction time is therefore very short, and results in a very small reactor. Due to the high length-to-diameter ratio of the channels, near plug-flow behavior is obtainable. This is beneficial for deep conversion as well as reduced by-product formation.

The advantages of milli-size channels over micro-channels are that plugging is less of an issue, even feed distribution over the channels is also more easily achieved, and the pressure drop is not very high. For gas flows, this means a less expensive compressor, and finally the manufacturing of milli-channel system costs much less. For instance, corrugated plates can be used to create the two channel systems [84]. The precise channel diameter and construction material can be optimized for the specific reaction considered [79].

### 2.6.2.10 Future outlook on micro- and milli-reactors

Applications for micro and millireactor devices are still rare in the industry [78, 85, 86] and only a few commercial scale applications in the fine chemicals industries are mentioned [85, 87].

Information on design and manufacturing of micro devices for pharmaceuticals and fine chemicals is available in the public domain [77]. Also, many technology providers are available with specific knowledge on design, manufacturing the devices, and scale-up.

The design and manufacturing is also in development. Three-D printing is likely to become an important technique for these micro- and millireactor devices [88]. In particular, the freedom of choice of construction material, the optimized structure for mixing, heat transfer and mass transfer and narrow RTD could mean that advanced lab-scale and larger scale devices can be manufactured quickly. The use of ceramic material in combination with 3-D printing is particularly promising. Heat transfer rates will be high and corrosion problems low [78]. Also, scale-up, without the empirical step by step scale-up, is very attractive. This will shorten the time to market, which is the overriding economic factor for new pharmaceutical, fine chemical, and specialty chemical products. Kleiner indicated that the shorter time to market using the modular scaled milli-reactor system for a specialty polymer had a Net Present Value of 100 M€ [89]. A very interesting commercial-scale milli-reactor process for gas-to-liquid application is described by Baxter. [84].

Hence, it is likely that micro and milli-channel devices will be applied on commercial scale for pharmaceuticals, fine and specialty chemicals, and also for other smaller capacity-scale applications.

### 2.6.2.11 Impinging streams for mixing

In the 1980s, I noticed the use of impinging jets to achieve very rapid mixing of gases (milliseconds mixing time) for a commercial-scale application in my company. Scale-up knowledge was obtained by pilot plants, modeling and experimental validation by cold flow tests.

Nowadays, a lot of information is available in the form of open literature. Impinging jet streams in the turbulent flow regime cause very fast mixing at both micro and meso scale [90]. Even if the viscosity between the two jet flows is different, fast mixing down to the molecular level can still be achieved [91].

The impinging jet mixing may be integrated with a reactor. Also, the impinging jet mixings may be combined with crystallization. If, for instance, one jet is an anti-solvent and the other jet contains a substance to be crystallized, then, due to the rapid mixing, only small uniform particles are formed [92].

Information on the design of impinging jet mixing is provided by Kresta [93]. Scale-up for single phase mixing can be model-based using computational fluid dynamics. The model has to be experimentally validated. This can be by a small-scale pilot plant (reactor) with real feeds and by a scale cold-flow model with the same dimensions as the commercial scale. Scale-up for crystallization is best carried out by model-supported pilot plant tests at two scales.

### 2.6.2.12 Compact plate heat exchangers

In compact plate heat exchangers, the heat transfer walls are plates and not tubes. The spaces between the plates are narrow. These two geometrical features create a high mass transfer area per volume as well as high heat transfer coefficients. This, in turn, means that high heat transfer rates per unit volume are obtainable at low temperature differences. The latter means that primary energy is saved compared to tube-and-shell heat exchangers [94].

For a long time, these devices were not used on a large scale in the process industries due to unreliable construction that resulted in leakage and feed maldistributions. Yaici provides information on maldistribution occurrence depending on inlet configuration [95]. These problems have been recognized and can now be prevented by proper design and construction [96].

For single phase Newtonian fluids, scale-up to commercial scale can be done by proven design models. For non-Newtonian fluids, scale-up can be done by a CFD model validated experimentally by a test unit with the same geometry, same fluid, and the same temperatures.

### 2.6.2.13 Membrane separations

It can be debated whether membrane separation belongs to intensified equipment. It can also be seen as an established unit operation. Some information is provided here. The reader may refer to textbooks such as by Cui [97], to find more information. For commercial-scale implementation, many technology providers are available with design and scale-up information.

In membrane separations, one molecule type is transported through the membrane material to the other side to produce a pure stream called the permeate. All other molecules stay on the feed side of the membrane material. This stream is called the retentate. The driving force is a concentration difference and often, also the pressure difference between the retentate and permeate side of the membrane.

Membrane separations are very good in producing a very pure permeate product stream. However, they are not good in obtaining a high yield of that product component in the feed stream. With ever increasing yield, the required pressure difference increases exponentially until the physical limitation of the membrane is reached.

If a high yield of the permeate component is needed, then a hybrid combination of the membrane with a different separation technique should be considered. A combination of a membrane and distillation is often a good choice. The distillation continuously pre-concentrates the feed to the membrane so that a nearly 100% yield of permeate component can be obtained, without an excessive pressure difference.

When the retentate is the product stream, such as in a thickened sugar juice, and water is the permeate, then the yield can be 100%, as the feed stream is concentrated by separating some of the water.

The largest established membrane separations are liquid phase water separation from aqueous solutions called reverse osmosis, and gas phase hydrogen separation from a gas mixture containing carbon dioxide. Reverse osmosis is applied to produce drinking water from seawater and to produce thickened juices in the food industry. Hydrogen purification is a mainly applied in oil refining.

Design methods are available for complete membrane units with multiple tubes. Scale-up for new applications is done by the brute-force scale-up method, by testing in a pilot plant consisting of a single tube or a few tubes of commercial scale length, flow velocities, and feed quality. Commercial scale design is then achieved by numbering up the tubes.

## 2.7 Domain combination options

Breakthrough process designs need not fall in a single domain category. Combinations of categories can be very beneficial. An example of such a combination on the biomass pyrolysis process to stable liquid biofuel is provided in Chapter 13.

No design guidelines can be given for combining domains. The inventor may be inspired by this chapter to find his own promising solutions.

## 2.8 Exercises

### Exercise 2.8.1: Top manager asks engineer about PI advantages

A young engineer stands in the elevator next to his CEO. The CEO says: "I just read a headline in Chemical Engineering on Process Intensification. What are the advantages of applying process intensification inside our company?"

Question: What should the young engineer answer, before the CEO leaves the elevator at the top floor?

### Exercise 2.8.2: PI department in chemical company

A very large bulk chemical company wants to start a process-intensification department that can provide advice to process and product R&D people on PI. The new head of this department asks you, his first staff member, to provide a short list of main PI design methods.

Question: Which design methods would you include on the list?

### Exercise 2.8.3: Deaerating water at oil platform at sea

As a young process engineer of an Oil & Gas company in the upstream part of the company, you are asked by the head of the engineering department to select a technology for de-aerating seawater to be used as cooling water. He says that the smallest and lightest technology should be chosen. For de-airration, a gas-liquid contactor is needed.

Question: Which gas-liquid contactor should you choose?

### Exercise 2.8.4: Rapid heating of soja

An engineer in an animals feed production company is told to enhance the nutritional value of soja beans for pigs feed. The nutritionist explains to him that a certain protein in the soja beans acts as poison for the pig digestion system. This protein can be converted to harmless protein by heating it. It is being done by heating using steam. The process step by which other proteins are also converted to less nutritional value requires one minute.

Question: What heating up method that requires far less than one minute can be applied to heat the entire soja bean?

### Exercise 2.8.5: EPC company wants efficient equipment for mixing liquids

A young process engineer is given the task of finding a solution for mixing two liquids in an existing bulk chemicals process. The output stream is 100 kt/year. The plot area for the equipment is $3 \times 3$ m². The mixing should be perfect.

Question: What mixing equipment should be selected?

# References

[1]    Schaschke, CA, Dictionary of chemical engineering, OUP, Oxford, 2014.
[2]    Green, DW, Perry's Chemical Engineers' Handbook, 8th, McGraw-Hill, New York, 2008.
[3]    Keil, FJ, Process intensification, Reviews in Chemical Engineering, 2018, 34.2, 135–200.
[4]    Stankiewicz, A, Gerven, van T., and Stefanides, G, The fundamentals of process intensification, Wiley-VCH, Weinheim, 2019.
[5]    Harmsen, J, Haan, de AB, and Schwinkels, PLJ, Product and process design driving innovation, De Gruyter, Berlin, 2018.
[6]    Harmsen, J, Industrial process scale-up a practical innovation guide from idea to commercial implementation, 2nd, Elsevier, Amsterdam, 2019.
[7]    Nguyen Van Duc Long, Lee, Moonyong, Review of retrofitting distillation columns using thermally coupled distillation sequences and dividing wall columns to improve energy efficiency, Journal of chemical engineering of Japan, 47(2), 87–108. January 2014.
[8]    Rangaiah, GP, editor., Chemical process retrofitting and revamping: Techniques and applications, John Wiley & Sons, Hoboken, 2016.
[9]    Portha, JF, Falk, L, and Commenge, JM, Local and global process intensification, Chemical Engineering and Processing: Process Intensification, 2014, Oct 1 84, 1–3.
[10]   Anastas, PT, and Warner, JC, Green chemistry- theory and practice, Oxford University Press, Oxford, 1998.
[11]   Anastas, PT, and Williamson, TC, Green chemistry – frontiers in benign chemical synthesis and process, Oxford University Press, Oxford, 1998.
[12]   Hammer, SC, Knight, AM, and Arnold, FH, Design and evolution of enzymes for non-natural chemistry, Current, Opinion in Green and Sustainable Chemistry, 2017, Oct, 1(7), 23–30.
[13]   Arnold, FH, and Georgiou, G, editors., Directed enzyme evolution: Screening and selection methods, Springer Science & Business Media, 2003.
[14]   Kun, LY, editor., Microbial biotechnology: Principles and applications, World Scientific Publishing Company, 2003.
[15]   Siirola, JJ, and Rudd, DF., Computer-aided synthesis of chemical process designs, From reaction path data to the process task network. Industrial & Engineering Chemistry Fundamentals., 1971, Aug, 10(3), 353–362.
[16]   Rudd, DF, Powers, GJ, and Siirola, JJ, Process synthesis, Prentice-Hall, 1973.
[17]   Sauar, E, Energy efficient process design by equipartition of forces, PhD thesis, Northern University Trondheim, 1998.
[18]   Siirola, JJ An industrial perspective on process synthesis. In AIChE Symposium Series 1995 91, 304, 222–234. New York, NY: American Institute of Chemical Engineers, 1971-c2002.
[19]   Douglas, JM, Conceptual design of chemical processes, McGraw-Hill, New York,1988.
[20]   Seider, WD, et.al., Product and process design principles, John Wiley & Sons, Hoboken, 2010.

[21] Qi, Z, and Sundmacher, K., Geometrically locating azeotropes in ternary systems, Industrial & engineering chemistry research, 2005, 44(10), 3709–3719.

[22] Harmsen, G.J, Reactive distillation: The front-runner of industrial process intensification: A full review of commercial applications, research, scale-up, design and operation, Chemical Engineering and Processing: Process Intensification, 2007, 46(9), 774–780.

[23] Harmsen, J, 2013, Implementation of Process Intensification in Industry, Process Intensification for Green Chemistry, Boodhoo, KVK and Harvey, AP, Eds., J. Wiley, Chichester, 2013, 393–400.

[24] Harmsen, J 2017, Wikisheet Reactive Distillation, sourced 10-1-2020 https://www.rvo.nl/sites/default/files/2017/08/Wikisheet%20Reactive%20Distillation%20-%20RVO.pdf

[25] Luyben, WL, and Yu, C, Reactive distillation design and control, J. Wiley, Hoboken, 2008.

[26] Harmsen, GJ, and Chewter, LA, Industrial applications of multi-functional, multi-phase reactors, Chemical Engineering Science, 1999, 54, 1541–1545.

[27] Agreda, VH, Partin, LR, and Heise, WH, High-purity methyl acetate via reactive distillation, Chemical Engineering Progress, February 1990, 40–46.

[28] CDTECH, http://www.cdtech.com/indexset.htm?aboutus.htm accessed 9-2-2016.

[29] Seyfert, W, Upwind, trends and challenges in Process Engineering Research, plenary lecture. BASF Senior Vice President, ECCE10, Nice, 2015.

[30] Kiss, AA, Dividing-wall column, in advanced distillation technologies: Design, control and applications, John Wiley & Sons, Ltd, Chichester, UK, 2013.

[31] Schoenmakers, H, Dividing Wall Columns, presentation at PIN NL Meeting 17-6-2010, http://www.rvo.nl/file/2507, accessed 16.02.2016.

[32] Schultz, MA, et.al., Reducing costs with dividing-wall columns, Chemical Engineering and Processing, 2002, 64–71.

[33] Julius Montz GmbH, Catalogue, http://www.nt.ntnu.no/users/skoge/prost/proceedings/distillation10/DA2010%20Sponsor%20Information/Montz/MONTZ%20Main%20Cataloque.pdf accessed 17.02.2016

[34] Roza, M (Sulzer), The magic of Dividing Wall Columns, presentation 19 April 2012 at Spring Session PINNL, http://www.traxxys.com/downloads/Files/The%20Magic%20of%20Dividing%20Wall%20Columns%2019-4-12.pdf accessed 17.02.2016

[35] Till, A, et. al., Model predictive control of integrated unit operations: Control of a divided wall column, Chemical Engineering and Processing, 2004, 43, 347–355.

[36] Dejanović, L, and Olujić, MŽ, Dividing wall column – A breakthrough towards sustainable distilling Chem, Chemical Engineering and Processing, 2010, 49, 559–580.

[37] Bunimovich, GA, and Matros, YS, Presentation, Chemreactor-19 conf., Vienna, 2010

[38] Matros, YS, Bunimovich, GA, and Noskov, AS, Catalysis Today,17 (1993) 261–274.

[39] Silvester, PL, and Hudgins, RR, Periodic operation of chemical reactors, Butterworth-Heinemann, Oxford, 2012.

[40] Bunimovich, GA, and Matros, YS, Saving energy in regenerative oxidizers, Chemical Engineering, March 2010, 26–32.

[41] McNamara, D, et.al., Mission impossible, Hydrocarbon Engineering March (2006) 1–4.

[42] Matros technologies, http://www.matrostech.com accessed 15.02.2016.

[43] Gerhart Eigenberger*, Grigorios Kolios1, Ulrich Nieken, Thermal pattern formation and process intensification in chemical reaction engineering, Chemical Engineering Science, 62, 2007, 4825–4841.

[44] Marin, MP, et.al., Simplified design methods of reverse flow catalytic combustors for the treatment of lean hydrocarbon–air mixtures, Chemical Engineering and Processing: Process Intensification, 2009, 48, 229–238.

[45] Zagoruiko, AN, The reverse-flow operation of catalytic reactors: History and prospects, Catalysis, 10, 2012.

[46] Bos, ANR, Kabra, GR, Lange, JP, and Co, Shell Oil, 2010. Process for removing contaminants using reverse flow reactor with integrated separation. U.S. Patent 7,763, 174.

[47] Isozaki, C, Katagiri, T, Nakamura, Y et.al. in: "Unsteady State Processes in Catalysis, Proc. of the Int. Conf. 5–8 June, 1990, Novosibirsk, USSR, VSP, Utrecht-Tokyo, 1990, 637.

[48] McDonough, JR, Phan, AN, and Harvey, AP, Rapid process development using oscillatory baffled mesoreactors – a state-of-the-art review, Chemical Engineering Journal, 2015, 265, 110–121.

[49] Regier, M, Knoerzer, K, and Schubert, H, eds., The microwave processing of foods, Woodhead Publishing, Stankiewicz, 2016, 20190613, 173.

[50] Visscher, F, Van der Schaaf, J, Nijhuis, TA, and Schouten, JC, Rotating reactors–a review, Chemical Engineering Research and Design, 2013, Oct 1, 91(10), 1923–1940.

[51] GiteshDubal, Process Engineer, B&P Process Equipment and Systems LLC, 1000 Hess Ave., Saginaw, MI 48601, Podbielniak contactor – a unique liquid-liquid extractor, Pharmaceutical online. Sourced 9 April 2019, https://www.pharmaceuticalonline.com/doc/podbielniak-contactor-a-unique-liquid-liquid-0003

[52] Doble, M, and Kruthiventi, AK., Green chemistry and processes, Academic,, 2007.

[53] Oxley, P, Brechtelsbauer, C, Ricard, F, Lewis, N, and Ramshaw, C, Evaluation of spinning disk reactor technology for the manufacture of pharmaceuticals, Industrial & Engineering Chemistry Research, 2000, Jul 3, 39(7), 2175–2182.

[54] Reay, D, Ramshaw, C, and Harvey, A, Process intensification: Engineering for efficiency, sustainability and flexibility, Butterworth-Heinemann, 2013 Jun 5.

[55] Brechtelsbauer, C, Lewis, N, Oxley, P, Ricard, F, and Ramshaw, C, Evaluation of a spinning disc reactor for continuous processing 1, Organic Process Research & Development, 2001, Jan 19, 5(1), 65–68.

[56] Pask, SD, Nuyken, O, and Cai, Z, The spinning disk reactor: An example of a process intensification technology for polymers and particles, Polymer Chemistry, 2012, 3(10), 2698–2707.

[57] Meeuwse, M, Lempers, S, van der Schaaf, J, and Schouten, JC, Liquid – solid mass transfer and reaction in a rotor–stator spinning disc reactor, Industrial & Engineering Chemistry Research, 2010, May 11, 49(21), 10751–10757.

[58] Manzano Martînez, AN, van Eeten, KM, and Schouten, JC, van der Schaaf J. Micromixing in a Rotor–Stator Spinning Disc Reactor, Industrial & engineering chemistry research, 2017, Aug 30, 56(45), 13454–13460.

[59] Meeuwse, M, van der Schaaf, J, and Schouten, JC, Multistage rotor-stator spinning disc reactor, AIChE Journal, 2012, Jan, 58(1), 247–255.

[60] Visscher, F, Van Der Schaaf, J, De Croon, MH, and Schouten, JC., Liquid–liquid mass transfer in a rotor–stator spinning disc reactor, Chemical Engineering Journal, 2012, Mar, 15(185), 267–273.

[61] Van Der Schaaf, J, Visscher, F, Bindraban, D, and Schouten, JC, inventors. Device for multiphase and single phase contacting. World patent WO 2012150226. 2015 Jan 28.

[62] Visscher, F, Nijhuis, RT, de Croon, MH, van der Schaaf, J, and Schouten, JC, Liquid–liquid flow in an impeller–stator spinning disc reactor, Chemical Engineering and Processing: Process Intensification, 2013, Sep 1 71, 107–114.

[63] Wang, Z, Yang, T, Liu, Z, Wang, S, Gao, Y, and Wu, M, Review and analysis of mass transfer process in rotating packed bed, Chemical Engineering and Processing-Process Intensification, 2019, Mar 30.

[64] Wenzel, D, Gerdes, N, Steinbrink, M, Ojeda, LS, and Górak, A, Liquid distribution and mixing in rotating packed beds, Industrial & Engineering Chemistry Research, 2018 Nov 1.

[65] Yang, YC, Ouyang, Y, Zhang, N, Yu, QJ, and Arowo, M, A review on computational fluid dynamic simulation for rotating packed beds, Journal of Chemical Technology & Biotechnology, 2019, Apr, 94(4), 1017–1031.

[66] Guo, J, Jiao, W, Qi, G, Yuan, Z, and Liu, Y, Applications of high-gravity technologies in gas purifications: A review, Chinese Journal of Chemical Engineering.2019, Jan 26.

[67] Trent, DL, Chemical processing in high gravity fields, in Re-engineering the Chemical Processing Plant, ed., Stankiewicz, A and Moulijn, JA, Marcel Dekker, New York, 2004, 33–68.

[68] Fang, JZ, and Lee, DJ, Micromixing efficiency in static mixer, Chemical Engineering Science, 2001, Jun 1;, 56(12), 3797–3802.

[69] Brechtelsbauer, C, and Ricard, F, Reaction engineering evaluation and utilization of static mixer technology for the synthesis of pharmaceuticals, Organic process research & development, 2001, Nov 16, 5(6), 646–651.

[70] Thakur, RK, Vial, C, Nigam, KD, Nauman, EB, and Djelveh, G, Static mixers in the process industries – a review, Chemical Engineering Research and Design, 2003, Aug 1, 81(7), 787–826.

[71] Wood, J, Monolith Reactors for Intensified Processing in Green Chemistry, Boodhoo, K and Harvey, A, Process Intensification for Green Chemistry, Wiley Chichester UK, 2013, 175–199.

[72] Devatine, A, Chaumat, H, Guillaume, S, Tchibouanga, BT, Martínez, FD, Julcour, C, and Billet, AM, Hydrodynamic study of a monolith-type reactor for intensification of gas-liquid applications, Chemical Engineering and Processing: Process Intensification, 2017, Dec, 1(122), 277–287.

[73] Haase, S, Murzin, DY, and Salmi, T, Review on hydrodynamics and mass transfer in mini-channel wall reactors with gas–liquid Taylor flow, Chemical Engineering Research and Design, 2016, Sep, 1(113), 304–329.

[74] Manfe, MM, Kulkarni, KS, and Kulkarni, AD, Industrial application of monolith catalysts/reactors, IJAERS, 2011, 1, 1–3.

[75] Lopes, JP, Rodrigues, AE, Önsan, ZI, Avci, AK, Wiley, John, Inc., Sons, and Hoboken, NJ, USA, 2016.

[76] Seyfert, W, Upwind, trends and challenges in Process Engineering Research, plenary lecture. BASF Senior Vice President, ECCE10, Nice, 2015.

[77] Wiles, C, and Watts, P, Micro reaction technology in organic synthesis, CRC Press, 2016.

[78] Reintjens, R, Ager, DJ, and De Vries, AH, Flow chemistry, how to bring it to industrial scale?, Chimica Oggi, 2015, Jul 1, 33(4), 21–24.

[79] Reintjens, R, and de Vries, A, Microreactors: Lessons Learned From Industrial Applications, Chemical Engineering, 2016, Aug 1, 123(8), 40.

[80] Harmsen, J, Economics and Environmental Impact of Process Intensification: An assessment for the petrochemical, pharmaceutical and fine chemicals industries, rocess Intensification for Green Chemistry, Boodhoo, KVK and Harvey, AP, Eds., John Wiley & Sons, Chichester, 2013, 369–378.

[81] Ainsworth, D, Planning and designing a pharmaceutical facility: A process designer's view, Pharmaceutical Technology Europe, 17(9), 2005.

[82] Jensen, KF, Flow chemistry – microreaction technology comes of age, AIChE Journal, 2017, Mar, 63(3), 858–869.

[83] Jimenez-Gonzales, AD, Fine Chemical Carbonyl Process, Glaxo Smith Kleine, Oral presentation AIChE Spring meeting 2001.

[84] Baxter, Iain, Modular GTL as an Offshore Associated Gas Solution, presentation at Deep offshore technology Int. Amsterdam 2010, http://www.compactgtl.com/wp-content/documents/110119_CompGTL_12p_A4_HR_FINAL.pdf accessed 17.02.2016.

[85] Schwalbe, T, et.al., Chemical synthesis in microreactors, microstructured reactor systems, CHIMIA, 2002, 2002, 56(11), 636–646.

[86] Poechlauer, P, Manley, J, Broxterman, R, Gregertsen, B, and Ridemark, M, Continuous processing in the manufacture of active pharmaceutical ingredients and finished dosage forms: A industry perspective, Organic Process Research & Development, 2012, Oct 10, 16(10), 1586–1590.

[87] Schnider, C, 2018, Lonza, From Batch to continuous Fluorination, Presentation at Process Intensification Network, Netherlands meeting, at DIFFER, Eindhoven, 27th of June.

[88] Potdar, A, Thomassen, LC, and Kuhn, S, Scalability of 3D printed structured porous milli-scale reactors, Chemical Engineering Journal, 2019, May 1 363, 337–348.

[89] Kleiner, M, 2011, Smart Production by modularisation, Oral presentation, ECCE8, Berlin, 2011.

[90] Hao, YU, Seo, JH, Hu, Y, Mao, HQ, and Mittal, R, Flow physics and enhanced mixing in confined impinging jet mixers, Bulletin of the American Physical Society, 2018 Nov 20.

[91] Brito, MS, Esteves, LP, Fonte, CP, Dias, MM, Lopes, JC, and Santos, RJ., Mixing of fluids with dissimilar viscosities in Confined Impinging Jets, Chemical Engineering Research and Design, 2018, Jun, 1(134), 392–404.

[92] Jiang, M, Li, YE, Tung, HH, and Braatz, RD, Effect of jet velocity on crystal size distribution from antisolvent and cooling crystallizations in a dual impinging jet mixer, Chemical Engineering and Processing: Process Intensification, 2015, Nov 1 97, 242–247.

[93] Kresta, SM, Etchells III, AW, Atiemo-Obeng, VA, and Dickey, DS, editors., Advances in industrial mixing : a companion to the handbook of industrial mixing, John Wiley & Sons, 2015 Nov 16.

[94] Hesselgreaves, JE, Law, R, and Reay, D, Compact heat exchangers: Selection, design and operation, Butterworth-Heinemann, 2016 Sep 26.

[95] Yaïci, W, Ghorab, M, and Entchev, E, 3D CFD study of the effect of inlet air flow maldistribution on plate-fin-tube heat exchanger design and thermal–hydraulic performance, International Journal of Heat and Mass Transfer, 2016, Oct, 1(101), 527–541.

[96] Kahn, E, Innovate or perish: Managing the enduring technology company in the global market, John Wiley & Sons, Hobooken, 2007.

[97] Cui, ZF, and Muralidhara, HS, Membrane technology: A practical guide to membrane technology and applications in food and bioprocessing, Elsevier, 2010 Sep 23.

[98] Visscher, F, de Hullu, J, de Croon, MH, van der Schaaf, J, and Schouten, JC, Residence time distribution in a single-phase rotor–stator spinning disk reactor, AIChE Journal, 2013, Jul, 59(7), 2686–2693.

# 3 The role of sustainable development goals

## 3.1 Introduction

In September 2015, the United Nations General Assembly met in New York to adopt *The 2030 agenda for sustainable development* to make our world more human, sustainable, prosperous, and peaceful. Among others, the World Business Council for Sustainable Development was also involved in developing this agenda and agreed with the end product: *The 2030 agenda* [1].

*The 2030 agenda* is an ambitious plan of action. Its main objectives are to free the human race from the tyranny of poverty and want, and to heal and secure our planet. It is stated that bold and transformative steps are urgently needed to shift the world onto a sustainable and resilient path. The member states of the United Nations promise that this ambition plan of action is a collective journey and that no member will be left behind.

*The 2030 agenda* specifies 17 sustainable development goals (SDGs). Each SDG has specified subgoals. Every SDG contributes to realizing a more human, prosperous, and peaceful world.

*The 2030 agenda* states that global partnership is required to realize this agenda. It also invites private businesses to contribute to the realization of this agenda. It states in article 67:

> Private business activity, investment and innovation are major drivers of productivity, inclusive economic growth and job creation. We acknowledge the diversity of the private sector, ranging from micro-enterprises to cooperatives to multinationals. We call upon all businesses to apply their creativity and innovation to solving sustainable development challenges. We will foster a dynamic and well-functioning business sector, while protecting labor rights and environmental and health standards in accordance with relevant international standards and agreements and other ongoing initiatives in this regard.

In general, the top management of international companies is familiar with the SDGs and want to contribute actively to the realization of its targets. The most important means by which the top management can to contribute to the realization of SDGs are by the vision and strategy process, the business planning process, the purchase policy, the sustainable development policy, and allocation decisions. Process intensification (PI) can contribute strongly to the realization of SDGs. PI requires a renewal of the existing industrial innovation practices and the existing industrial infrastructures and networks. It is about a radical cultural change.

Usually, engineers who are active in PI have only a general knowledge of SDGs. Mostly they are familiar with some characteristic goals like the reduction of poverty and the development of a sustainable world. However, they are not familiar with the structure and the setup of the SDGs and do not have the knowledge to contribute to the realization of SDGs in process innovation.

https://doi.org/10.1515/9783110657357-003

It goes without saying that only the process industries can contribute to the realization of SDGs when top management defines their policy with respect to the SDGs and when engineers can make this policy concrete in the development and implementation of intensified processes. It should be noted that the relationship between policy of the top management and the subsequent concretization by engineers is not a top-down process. Intensive dialogue between the top management and engineers is needed to define workable policies and to develop fruitful concretizations.

The objective of Section 3.2 is to discuss the SDGs from the perspective of the top management. In Section 3.3, the most important SDGs are discussed from the perspective of the engineers. In Section 3.4, we present one SDG in more detail. In Section 3.5, we describe how the SDGs can be applied in the different stages of a PI project.

## 3.2 SDGs: the perspective of top management

Process intensification, by definition, can contribute to delivering on SDGs: that is, by innovation and development of processes that have better raw material and/or energy efficiency of an order of magnitude than conventional processes. PI processes are therefore putting less stress on our natural resources.

The United Nations have defined 17 SDG. Each goal is elaborated in detail. The titles of these goals are quite self-explanatory.

SDG 1: *End poverty in all its forms everywhere*

SDG 2: *End hunger, achieve food security and improved nutrition, and promote sustainable agriculture*

SDG 3: *Ensure healthy lives and promote well-being for all, at all ages*

SDG 4: *Ensure inclusive and equitable quality education and promote lifelong learning opportunities for all*

SDG 5: *Achieve gender equality and empower all women and girls*

SDG 6: *Ensure availability and sustainable management of water and sanitation for all*

SDG 7: *Ensure access to affordable, reliable, sustainable, and modern energy for all*

SDG 8: *Promote sustained, inclusive and sustainable economic growth, full and productive employment and decent work for all*

SDG 9: *Build resilient infrastructure, promote inclusive and sustainable industrialization, and foster innovation*

SDG 10: *Reduce inequality within and among countries*

SDG 11: *Make cities and human settlements inclusive, safe, resilient, and sustainable*

SDG 12: *Ensure sustainable consumption and production patterns*

SDG 13: *Take urgent action to combat climate change and its impacts*

SDG 14:   *Conserve and sustainably use the oceans, seas and marine resources for sustainable development.*
SDG 15:   *Protect, restore, and promote sustainable use of terrestrial ecosystems, sustainably manage forests, combat desertification, and halt and reverse land degradation and halt biodiversity loss.*
SDG 16:   *Promote peaceful and inclusive societies for sustainable development, provide access to justice for all and build effective, accountable and inclusive institutions at all levels.*
SDG 17:   *Strengthen the means of implementation and revitalize the Global Partnership for Sustainable Development*

First of all, it has to be noted that a couple of SDGs influence the policy and strategy of the process industry directly. For example, SDG 7 is about an affordable, reliable, sustainable, and modern energy, SDG 8 is about an inclusive and sustainable economic growth and decent work for all, SDG 9 is about a resilient infrastructure that promotes sustainable industrialization and fosters innovation, SDG 12 is about sustainable consumption and production patterns, and SDG 13 is about urgent action to combat climate change and its impacts.

Second, some SDGs are important for specific process industries. For example, SDG 2 is about ending hunger and achieving food security which is important for industries in the field of food and agriculture, SDG 3 is about healthy lives and promoting well-being which challenges industries in the field of food, agriculture, water production, and pharmacy, SDG 6 is about the availability of water and sanitation which is relevant for industries in the field of water production and water treatment, and SDG 11 is about inclusive, safe, resilient, and sustainable cities and human settlements that address, for example, industries in the field of building and road materials.

Third, some SDGs are relevant for international companies with subsidiaries in developing countries or for international companies that intend to allocate one of their plants in a developing country. For example, SDG 4 is about education and lifelong learning, SDG 5 is about gender equality, SDG 8 is about promoting sustainable economic growth and decent work for all, and SDG 10 is about reducing inequality within and among countries.

Finally, some SDGs have to be addressed mainly by governments and global political institutions. For example, SDG 1 is about ending poverty, SDG 2 is about ending hunger, achieving food security and promoting sustainable agriculture, SDG 3 is about ensuring healthy lives and promoting well-being, SDG 4 is about ensuring quality education and promoting lifelong learning, SDG 5 is about achieving gender equality, SDG 6 is about ensuring availability of water and sanitation for all, SDG 10 is about reducing inequality within and among countries, SDG 11 is about making cities and human settlements inclusive, safe, resilient and sustainable, SDG 14 is about conserving and sustainable use of oceans, seas, and marine resources, SDG 15 is about protecting, restoring, and promoting sustainable use of terrestrial

ecosystems, and halting biodiversity loss, SDG 16 is about promoting peaceful, inclusive and just societies, and SDG 17 is about strengthening and revitalization of
Global Partnership for Sustainable Development.

A successful engagement with the SDGs can be done by executing the following
steps [2, p. 22]:

1)  "Agree which SDGs your business and its value chain have an impact on, di-
    rectly and indirectly, in the countries in which you operate;
2)  Agree on a methodology and measure your business impact across all these
    SDGs;
3)  Understand where your business has a positive or negative impact on each SDG
    (businesses should try and examine all of the SDGs initially and then act upon
    and measure those most relevant to them);
4)  Understand the priorities of the governments your business operates under;
5)  Prioritize reducing negative impacts and increasing positive impacts according
    to what needs to be achieved by governments;
6)  Incorporate this learning into business planning and strategy;
7)  Evidence how you impact on the SDGs and your contribution."

The importance of each SDG for the industry has been described extensively and
illustrated with relevant showcases [2, 3].

## 3.3  SDGs: the perspective of engineers

In the foregoing section we have presented the 17 SDGs of the United Nations. We
have shown that these SDGs can be categorized into different categories. Some
SDGs have a direct influence on engineers who work on PI and other SDGs have
only an indirect influence [2, 3]. First, we will discuss the SDGs that are directly affected by PI: SDG 7, SDG 9, SDG 12, and SDG 13. Later, we will present some specific
examples with respect to SDG 2, SDG 3, and SDG 6.

### 3.3.1  Highlighting the most relevant SDGs

**SDG 7: Ensure access to affordable, reliable, sustainable, and modern energy for all**
PI, first and for all, contributes to a sustainable world by developing processes with
an energy efficiency that is of an order of magnitude better than conventional processes. Chapters 13, 14, and 15 show industrial cases that contribute to this goal.

This goal invites engineers to develop sustainable energy sources for PI processes. One example is to replace natural gas by hydrogen. Hydrogen can be generated by hydrolysis using electricity generated by solar panels. Another example is to

turn waste into fuel. It should be noted that this type of developments can also contribute to the development of affordable, reliable, and sustainable energy for all.

Finally, this goal challenges engineers working in the field of generation, transport, distribution, and storage of energy to apply design methods that have been developed for PI (see Chapter 2).

### SDG 9: Build resilient infrastructure, promote inclusive and sustainable industrialization, and foster innovation

PI requires a radical change in industrial innovation practices. It cannot be achieved by the traditional method of continuous improvement. It involves nothing less than new concepts, new ways of cooperation, and new ways of acting. In other words, it requires an innovation of industrial innovation practices itself. In addition, it also involves innovation in industrial infrastructure and networks. Leadership in chemical industry should create an excellent climate for out-of-the-box thinking and innovation to address the SDGs effectively. Engineers play a key role in developing such a climate. Cases described in Chapter 13, 14, and 16 show cases where engineers played this key role.

We would like to note that an inclusive and sustainable industrialization can be realized only by innovating our industrial innovation practices and innovating our industrial infrastructures and networks. There is no other route to realize this goal.

### SDG 12: Ensure sustainable consumption and production patterns

PI strongly contributes to this SDG because its processes have a material and/or energy efficiency that is an order of magnitude better than conventional processes. It has to be noticed that PI as such does not contribute to sustainable consumption patterns. However, by combining PI principles with the VIB model (chapter 4) and the ideas of circular economy [4, 5], PI will also contribute to sustainable consumption patterns.

PI can address some of the major challenges of the process industry to ensure sustainable production patterns. A well-known example is the industrial production of methyl acetate by Eastman company by integrating reaction, distillation, and extraction in a single unit operation, strongly reducing the energy consumption as described in chapter 14.

### SDG 13: Take urgent action to combat climate change and its impacts

Climate change is one of the world's pressing challenges. Globally, humans release over 36 billion tons of carbon dioxide per year which continues to increase [6]. The chemical industry accounts for around 7% of the global greenhouse gas (GHG) emissions or around 20% of the total industrial GHG emissions. Among the thousands of chemicals, only 18 of them account for 75% of the GHG emissions of the chemical industry. Global demand of these 18 products is forecast to increase by 200% by 2050 [7]. These data illustrate the challenge for the process industry to realize a significant net reduction of carbon emissions. Recycling and chemicals produced from renewable carbon sources are essential to reduce GHG emissions.

PI strongly contributes to reduce the carbon footprint of the process industry by strong reductions in the use of energy. PI can also contribute to this goal by intensification of the production of renewable carbon sources. For these reasons, PI should be the rule and not the exception when designing new plants or when redesigning existing plants.

### 3.3.2 Highlighting some specific SDGs

Some SDGs are important for specific process industries. We give some examples from the fields of food, pharmacy, and water.

SDG 2 is about ending hunger and achieving food security. PI in the field of food may contribute to this goal. For example, harvested food may be processed quickly and locally so that it becomes transportable and has a long shelf live, as in the case of cassava.

SDG 3 is about healthy lives and promoting well-being. PI in the field of pharmacy may contribute to this goal as shown in chapter 17.

SDG 6 is about the availability and sustainable management of water. Engineers are at the forefront of developing novel intensified processes for water management and local water production in remote areas. PI leads to novel process technologies to achieve significant (order of magnitude) size reduction in individual unit operations or the complete removal of process steps by performing multiple functions in fewer steps. This should lead to significant reductions in capital and running costs and improvements in process efficiency and safety [8]. Membrane distillation is a process intensified technology that enables selective separation of water vapor from seawater or contaminated groundwater streams to produce fresh drinking water. Membrane distillation can utilize low temperature solar heat to drive the process. It can be built in small modular units, enabling remote operation in areas where no or few utilities are available [9].

## 3.4 Detailed analysis of sustainable development goals: example of SDG 12

The goal of SDG 12 is to ensure sustainable consumption and production patterns. The second part of the goal – sustainable production patterns – lies at the heart of the process industry. However, this SDG requires more to be done. It also invites the process industry to contribute to the first part of the goal – namely, sustainable consumption patterns. At first sight, top management and engineers may believe that the first part of this goal is outside their circle of influence. At second sight, however, a more nuanced opinion in possible. One does not have to be a prophet to predict that in the future it will become increasingly expensive to dispose of used materials. At the same time, the pressure on the industry to think circularly will increase. All this is a threat, but also a business opportunity. The idea of circular material flows can result in both a reduction in material costs as well as new business opportunities. Additionally, psychologists are developing a variety of nudging technologies to support consumers and companies to realize more sustainable and more circular material flows.

*The 2030 agenda* elaborates the goal to ensure sustainable consumption and production patterns as follows:

12.1 Implement the 10-Year Framework of Programs on Sustainable Consumption and Production Patterns, all countries acting, with developed countries taking the lead, considering the development and capabilities of developing countries.

12.2 By 2030, achieve sustainable management and an efficient use of natural resources.

12.3 By 2030, halve the per capita global food waste at the retail and consumer levels and reduce food losses along production and supply chains, including post-harvest losses.

12.4 By 2020, achieve the environmentally sound management of chemicals and all wastes throughout their life cycle, in accordance with agreed international frameworks, and significantly reduce their release to air, water, and soil in order to minimize their adverse impacts on human health and the environment.

12.5 By 2030, substantially reduce waste generation through prevention, reduction, recycling and reuse.

12.6 Encourage companies, especially large and transnational companies, to adopt sustainable practices and to integrate sustainability information into their reporting cycle.

12.7 Promote public procurement practices that are sustainable, in accordance with national policies and priorities.

12.8 By 2030, ensure that people everywhere have the relevant information and awareness for sustainable development and lifestyles in harmony with nature.

12.a Support developing countries to strengthen their scientific and technological capacities to move towards more sustainable patterns of consumption and production.

12.b Develop and implement tools to monitor sustainable development impacts for sustainable tourism that creates jobs and promotes local culture and products.

12.c Rationalize inefficient fossil-fuel subsidies that encourage wasteful consumption by removing market distortions, in accordance with national circumstances, including by restructuring taxation and phasing out those harmful subsidies, where they exist, to reflect their environmental impacts, taking fully into account the specific needs and conditions of developing countries and minimizing the possible adverse impacts on their development in a manner that protects the poor and the affected communities.

It goes without saying that these goals have to be discussed in detail to understand their importance for the process industry in general and PI in particular.

## 3.5 SDGs and stage-gate process of innovation

In this book we distinguish the following stages:
1) Discovery stage
2) Concept stage
3) Feasibility stage
4) Development stage
5) Engineering procurement and construction stage
6) Implementation stage

For each of these stages, the following questions have to be addressed:
1) How can PI contribute to the realization of SDGs?
2) What opportunities do SDGs offer for the process-intensified industry?

Each stage has its own objectives and dynamics. In chapters 6–11, we describe how the SDGs must be discussed and specified for every phase.

## 3.6 Epilogue

It is possible that the first reaction to the considerations presented above will be a deep sigh, reflecting that it is a lot of work to consider the SDGs. This is true. Yes, it takes a lot of time and effort to make sure that the new process contributes to the SDGs. However, the flip side of the coin is that carrying out this design and development with the SDGs in mind gives a lot of satisfaction. Deep down, every top manager and every engineer is motivated to contribute to a more human, sustainable, prosperous, and peaceful world. After all, we are not only discussing the world in which we will live but we are also discussing the world of our children and grandchildren.

## References

[1]   United Nations, 2015, Transforming our world: the 2030 agenda for sustainable development.
[2]   PwC, Gemis, and UNIDO, 2017, Delivering the Sustainable Development Goals. Seizing the opportunity in global manufacturing. Retrieved March 4 2020 from https://www.pwc.com/m1/en/publications/documents/delivering-sustainable-development-goals.pdf
[3]   United Nations and KPMG, 2016, SDG Industry Matrix. Industrial manufacturing. Retrieved March 4 2020 from https://www.unglobalcompact.org/docs/issues_doc/development/SDGMatrix-Manufacturing.pdf
[4]   European Commission, 2017, Circular Economy research and innovation – Connecting economic & environmental gains. Retrieved October 31, 2018 from https://ec.europa.eu/programmes/horizon2020/sites/horizon2020/files/ce_booklet.pdf
[5]   European Commission, 2018, A European Strategy for Plastics in a Circular Economy. Retrieved October 31 2018 from http://ec.europa.eu/environment/circular-economy/pdf/plastics-strategy-brochure.pdf
[6]   www.ourworldindata.org. Retrieved March 4, 2020.
[7]   www.globalefficiencyintel.com/new-blog/2018/chemical-industrys-energy-use-emissions. Retrieved March 4, 2020.
[8]   Coward, T, Tribe, H, and Harvey, AP, Opportunities for process intensification in the UK water industry: A review, *Journal of Water Process Engineering*, 21, February 2018, Pages 116–126. https://doi.org/10.1016/j.jwpe.2017.11.010
[9]   Alkhudhiri, A, Darwish, N, and Hilal, N, Membrane distillation: A comprehensive review, *Desalination*, 287, 15 February 2012, pages 2–18.

# 4 Values, interests, and beliefs – three perspectives on industrial innovation practices

## 4.1 Introduction

The meaning of the saying "fishes are bad hydrologists" is very obvious. Water is so natural for fishes that they do not reflect on the necessity and the complexity of their ecological system. In my experience (Maarten), this saying also holds for engineers in industrial innovation practices. They are fascinated by technology. They go all out to develop creative solutions. Innovation is so natural for them that they do not question the complexity of their innovation practice, until . . . Here, I would like to tell a story of an engineer who ran into problems.

The story is about Dr. Paulo Ribeiro, an eminent electrical engineer. He was a professor at the Calvin College, Grand Rapids, USA. Ribeiro's research topic was on electrical energy infrastructure of the future. This infrastructure will be more complex than the present one because it has to integrate sustainable energy resources, develop new distribution systems for customers with a range of consumption patterns, and implement smart control systems. During an extensive discussion, Ribeiro sighed: "It is impossible for an engineer to take the full complexity of these systems into account. I only have reduced models resulting in reduced designs that for their part result in sub-solutions and even wrong designs." This complaint resulted in the question: "Can philosophy support me to understand the complexity of this type of systems and to provide me to design better systems?" This question marked an intensive cooperation that resulted in two articles [1, 2]. During this cooperation, we learned a lot about industrial innovation practices. Step by step, we began to understand the world of the engineer.

First, we got an insight into the thinking and action of engineers. We discovered that they are driven by traditional values such as safety, efficiency, sustainability, and costs. Second, we found that engineers do not really understand the interests of their stakeholders. They divide stakeholders into three broad categories: electrical energy companies, customers, and authorities. They try to intuitively understand the interests of these stakeholders. Third, we found that engineers do not have an antenna for societal developments. For example, they do not fathom the importance of sustainable development goals of the United Nations. Based on our understandings, we proposed a model to overcome these shortcomings.

The encounter with Dr. Ribeiro challenged me as an engineer and as a philosopher. Can philosophy support engineers in understanding their innovation practices? Can it offer "tools" to facilitate the development of new products, processes, and infrastructures?

What is the challenge? Dr. Ribeiro's story is not an isolated one. I have had a couple of discussions with engineers who asked for my support. I would like to tell

https://doi.org/10.1515/9783110657357-004

two more stories to illustrate the necessity of practical models to understand innovation practices.

Dr. Joost van Hoof, an architect, wrote a thesis titled *Aging-in-place. The integrated design of housing facilities for people with dementia* [3]. In this thesis, he combined two existing models, The International Classification of Functioning, Disability and Health (ICF model) and the Model of Integrated Building Design (MIBD model). Essentially, the combination of these two models was already a breakthrough in thinking. However, this combination led to new questions. First, how do we know that the combination of two models leads to an "integral model"? Second, how do we relate the medical concepts of the ICF model to the building concepts of the MIBD model? The discussion about these two questions led to a long-term cooperation that resulted in an article in *Technology in Society* titled "Developing an integrated design model incorporating technology philosophy for the design of healthcare environments: A case analysis of facilities for psycho-geriatric and psychiatric care in The Netherlands" [4] and an article in the *Journal of Enabling Technologies* titled "A neurological and philosophical perspective on the design of environments and technology for older people with dementia" [5]. In this cooperation, we investigated the professional practice of an architect. We used different philosophical theories to understand the relationships between care, technology, and stakeholders.

Fred Holtkamp is an engineer who is active in the field of orthopedic engineering. One of his research topics is the quality of ankle foot orthoses (AFOs). An AFO is a piece of medical technology that is generally constructed of lightweight polypropylene-based plastic in the shape of the letter L with the upright part behind the calf or in front of the shank and the lower part running under the foot. They are attached to the calf with a strap and are made to fit inside accommodative shoes. This orthosis is intended to control the position and motion of the ankle. The reasons for the use of AFOs vary from congenital malformations to diseases and traumas such as poliomyelitis, multiple sclerosis, cerebrovascular accident (stroke), or head injuries resulting in an aberrant gait pattern. Holtkamp conducted a questionnaire study to obtain insights into the use of AFOs and the satisfaction of AFO users.

He found that one out of fifteen participants did not use the AFO at all. This result is striking because these people depended on a device for their daily functioning. Evidently, using this device generated more discomfort for them than not using it. Further, he showed that one out of four AFO users was dissatisfied with the device. Amongst other points, they complained about pain, discomfort, and hindrance. Finally, the investigators reported that even satisfied and very satisfied users mentioned serious problems in the daily use of the device.

These findings resulted in the question, "How to understand these results?" Later on, this question was reformulated as "Can philosophy help me to develop tools to improve the use of and satisfaction with AFOs?" The investigation of these questions resulted in the articles, "Professional Practices and User Practices: An Explorative Study in Health Care" [6] and "Understanding User Practices When

Drawing up Requirements – The Case of Designing Assistive Devices for Mobility" [7]. In this study, we applied the so-called practice model, a philosophical approach to unravel the complexity of professional practices (orthopedic doctor and orthopedic engineer) to unknot the world of AFO users and to propose a methodology to improve the quality of this type of assisted device.

The lessons from these stories are that engineers need adequate models to understand the complexity of their practice and they need methods to design better products, processes, and infrastructures. These models are provided in the next sections.

## 4.2 Practice approaches

In this book, we present the practice approach to unravel the complexity of industrial innovation practices. We are convinced that engineers can benefit from this approach. The main reason for us having so much faith in this approach is that it has been developed successfully by me and by engineers in various other fields. I have had a number of discussions with Paulo Ribeiro, Joost van Hoof, Fred Holtkamp, and other engineers about their innovation practices. On the one hand, I have proposed philosophical theories and the corresponding terminology. I have driven them crazy by asking the question "Please, explain?" again and again. On the other hand, they challenged me to leave philosophical terminology in favor of engineering clarity. They made me mad with their comments such as "It is too complex" or "Engineers do not think in this way."

The dialogues with engineers resulted in building blocks for a practice approach that fits into the world of an engineer. The main reasons are as follows:

### Complexity
Practice approaches offer methods to analyze and understand the complexity of technological products, technological installations, and technological networks. Additionally, they provide a vocabulary to characterize the nature of the different relationships between products, processes, and infrastructures.

### Professional practices
Practice approaches offer methods to understand the different practices in which engineers work, for example, practices of research & development (R&D), operations, marketing & sales, and that of chief executives. Moreover, they give more insight in the substantive contribution of different engineering practices in industrial innovation practices.

**Stakeholders**
Practice approaches also present methods to understand the nature of the different stakeholders of a chemical plant, for example, the practices of suppliers, customers, authorities, local residents, action groups, and so on. Specifically, they offer insights into the interests of these stakeholders and the approach to meet their interests in the design of the process installations.

**Human phenomena**
Practice approaches leave space for human phenomena like initiative, creativity, conflict, power, deceit, and so on. Practice approaches show that these types of phenomena are relevant and co-determine the performance of industrial innovation practices. Especially, these phenomena have a strong influence on the quality of the design process.

**Human-machine interaction**
Practice approaches stress that in all these practices, both human actors as well as technology play an important role. On the one hand, humans design and operate machines, and on the other hand, machines determine how humans have to act. It goes without saying that human–machine interactions are key to design safe installations and to realize operational excellence (see later).

**Operational excellence**
Practice approaches acknowledge the idea that professional practices have to operate according to "standards of excellence." These standards of operational excellence can only be realized by the design of controlled processes, an outstanding training of operators, a philosophy of continuous improvement, and an intense cooperation between operators, maintenance technicians, quality engineers, factory engineers, and (middle) management.

Our choice to opt for a practice approach is hence not out-of-the-blue. In organizational sciences, sociology, and philosophy, a tendency to research human practices is clearly visible [8–11]. One of the main reasons is that practice theories can "offer a radically new way of understanding social and organizational phenomena" [10].

## 4.3 The VIB practice model: values of engineers, interests of stakeholders, beliefs in society

The VIB practice model offers three perspectives to understand industrial innovation practices in process industries such as oil refinery, bulk chemicals, food processing, production of pharmaceuticals, specialty chemicals, and minerals refining.

It also provides guidelines for new innovations. It is argued that the complexity of these innovation practices can be understood and new innovations can be guided by applying three different perspectives:

V: Values of engineers

I: Interests of stakeholders

B: Beliefs in society

It has to be noted that these three perspectives are different points of view to investigate one and the same industrial innovation practice. It is similar to using three different pairs of glasses to investigate the same object. These perspectives partly highlight specific properties, partly reveal interactions, and partly show the same phenomena.

### 4.3.1 Values of engineers

Values of engineers refers to the world of engineers. This is about the "intrinsic values" of the process industry. Everyone will agree that process industries are involved in the production of materials. However, it is more than "just" production – it is about *controlled production*. The idea is that every hour, every day, every week, and every year, the process runs continuously and that the produced materials have the same quality. In Operations, all activities are focused on the control of the manufacturing process. The idea of control comes back to the technologies used, the design of the process, the use of techniques such as Statistical Process Control, the training of the process operators, the set-up of the maintenance schedules, and so on. We would like to mention that the processes for oil refinery and minerals refining, bulk chemicals and specialty chemicals, food processing and the production of pharmaceuticals, are quite different. However, all these processes have one characteristic in common: controlled production.

The intrinsic values specify the values that underlie the identity of a practice. All practices in the process industry share values such as process control, efficiency, effectivity, safety, and health. Depending on the specific process and the materials produced, these values have to be specified in detail. For example, let's take the value "health." In an oil refinery, this value refers mainly to the production process: the health of all employees involved, the health of citizens that live in the neighborhood of the plant, and the health of the product users. In the food industry, however, the the value "health" refers to not only the food production process but also to the consumption process. In other words, the food production process has to be designed in such a way that health problems in the consumption process are minimized. In the pharmaceuticals industry, the value "health" is again quite different. Specifically, it refers to the healing power of the product and the prevention of undesirable side effects of the treatment. Hence, the production process has to be designed in such a way that a decrease in the healing power or the probability of negative side effects is prevented. A similar analysis can be carried out for the other intrinsic values. Our conclusion is that every intrinsic value – that is, control, efficiency, effectivity, safety, and health – has its own specific "color" or "specific meaning" for each industrial sector.

We would like to make two remarks in addition to those mentioned in the previous section. First, it has to be noted that the idea of "controlled production" and "intrinsic values" are closely related. They can be described as two sides of the same coin. On the one hand, the intrinsic values provide additional specification to the idea of controlled production. They describe the nature of controlled production for a specific industrial practice. On the other hand, the idea of controlled production gives a further specification of the intrinsic values. It describes the context in which the intrinsic values have to be realized.

Second, engineers are experts in their own field. They have the feeling that they know what a controlled production is and which intrinsic values play a role. Often, their feeling is right. However, at times, it appears that their feelings need an update. This is especially so in view of the challenges of process intensification. The VIB practice model invites engineers to make their values explicit and to reflect on them.

In Section 4.4, we will show that the VIB model supports engineers to reflect on the present values in the company and define the values that are needed for developing intensified processes. That means, this approach is very suitable to make possible tensions between present values and future values explicit and to make them negotiable.

To identify the intrinsic values of an industrial company and to understand its specific color, the following method can be used:
1) Take as a starting point, the earlier mentioned values: control, efficiency, effectivity, safety, and health.
2) Brainstorm about values with engineers from different disciplines.
3) Study the vision and mission statements of the company with respect to the values.
4) Investigate the ethical codes of relevant engineering associations about the values.
5) Use the results of 1) to 4) to make a final list of engineering values.
6) Describe the specific meaning ("color") of every value for the practice.

Exercises as these are beneficial to the company because they invite engineers to reflect upon their own values and the values of the company. Values can be interpreted as the "invisible hand" that guides all employees in their actions. Jim Collins and Jerry Porras have shown in their book *Built to Last: Successful Habits of Visionary Companies* that value-driven companies are more sustainable and more profitable [12].

## 4.3.2 Interests of stakeholders

Interests of stakeholders is a perspective that focuses on the interests of external stakeholders. We would like to bring to attention that there may be a difference between the external stakeholders of the company and the external stakeholders of the industrial innovation practices (R&D department). If that's the case, both types of stakeholders have to be involved in the analysis. The most important stakeholders are suppliers of materials and technology, customers, local authorities, national authorities, shareholders, knowledge institutes, the local community, social action groups, and so on. Stakeholders are not "just" stakeholders. Every stakeholder has an own value or "color" and specific interests. The analysis of the interests of stakeholders can be carried out in three steps:

1)  Identify the relevant stakeholders of the present process and the innovated process.
2)  Investigate the specific quality of the interests of every stakeholder.
3)  Identify the strategic importance of every stakeholder.

### Step 1: Identification of stakeholders

Bucholtz and Carroll [13] offer an extensive list of stakeholders of a corporation from a business and societal perspective. Harmsen [14] provides a list of stakeholders in the process industry from an innovation perspective. These combined insights provide an overview of the most important categories of stakeholders that are affected by innovations such as process intensification. These categories are customers, suppliers of feedstock, suppliers of equipment, R&D institutions, competitors, shareholders, financial institutions, authorities, action groups, and local community.

It is very important to analyze both the stakeholders of the current process in the company and the stakeholders of the innovated process. Changes in stakeholder configurations have to be analyzed in detail to understand the (social) acceptance of the innovated process.

### Step 2: Investigate the specific quality of the interests of stakeholders

There are different types of stakeholders. For example, the nature of a research institution is quite different from that of a key customer. Additionally, the nature of a social action group cannot be compared with that of the shareholders. Every stakeholder has a specific nature or character: quality or qualifying function. For example, the quality of a research institution has to do with knowledge; its qualifying function is analytical. The quality of a key customer has to do with the purchase of goods; its qualifying function is economic. The quality of a social action group is addressing moral issues; its qualifying function is moral. In addition, the quality of shareholders is about return on investment; its qualifying function is economic.

The qualifying function not only describes the nature of a stakeholder but also defines its motives and offers a key to understand its behavior. For example, the objective of a national authority is to promote justice by making laws. Therefore, its qualifying function is juridical. This qualifying function not only describes the nature of the national authority (it is all about making and enforcing laws) but also describes its motives (to realize a just world) and its behavior (making laws and regulations, appointment of independent judges, checking compliance, investigating accidents).

The idea of the qualifying function also helps us to distinguish justified and unjustified interests. For example, it is a justified interest of national authorities to enforce that oil refineries comply with safety and environmental regulations, for example, by executing (unannounced) checks. However, it is an unjustified interest if authorities interfere with the technology of process intensification or the height of the dividend payment to shareholders. That's the responsibility of the company and not the business of a national authority. To give another example, it is a justified interest of the pharmaceutical industry to make a profit in order to reward shareholders and to develop new products. However, it is an unjustified interest when the pharmaceutical industry interferes with the prescriptions of medicines of doctors in hospitals to increase their profits. After all, the prescription of medicines is the responsibility of doctors and not that of the pharmaceutical industry.

The main stakeholders are:

### Commercial customers
Most customers are commercial companies. Their objective is to make a profit by adding value to society by their products and services. So, their main justified interest is their economic aspect.

### Noncommercial customers
Some customers are noncommercial organizations. For example, hospitals are important customers of the pharmaceutical industry. The main objective of a hospital is to care for their patients. The main justified interest is the moral aspect.

### Suppliers of feedstock
Suppliers of feedstock are commercial companies. Their objective is to run their business and make a profit. Their main justified interest is the economic aspect.

### Suppliers of equipment
Suppliers of equipment are commercial companies. Their objectives are to innovate, run their business, and make a profit. Their main justified interest is the economic aspect.

## R&D institutions

R&D institutions can be commercial or noncommercial. The objective of noncommercial R&D institutions is to acquire knowledge. Thus, their main justified interest is analytical. The objective of commercial R&D organizations is to make a profit by selling knowledge. Their main justified interests are economic and analytical.

## Competitors

Competitors are commercial organizations. They want to make a profit by adding value to society by selling products and services. Their main justified interest is economic.

## Shareholders

Shareholders primarily invest to make a profit, both by an increase in stock prices and by dividend payments. Their justified interests are economic. It should also be noted here that there is a significant proportion of shareholders investing primarily or even exclusively in "green" projects or sustainable businesses. They invest to make a profit but also because they want to contribute to a more sustainable world for the generations to come. Their justified interests are not only economic but also moral.

## Financial institutions

Financial institutions such as banks provide the chemical industry with financial services such as consultancy and loans. Their justified interests are primarily economic.

## Authorities

National and local authorities have the responsibility to enact laws and to enforce compliance with these laws. Their primary justified interest is juridical. It has to be noted that authorities – by enacting laws – also play an important role in the economy. Therefore, a secondary justified interest is economic.

## Action groups

Most action groups have an ideological background and have objectives in the field of justice, safety, health, pollution, and warming-up of the earth. There are differences in opinion on whether action groups have justified interests. It is generally believed that they have justified interests. Their main interest is morally qualified.

**Local community**
The local community comprises different parties, such as local residents, schools, tourist offices, local media, and so on. Their justified interests are quite diverse: economic, social, juridical, moral, and so on.

**Step 3: Identify the strategic importance of stakeholders**
Once stakeholders are identified and their justified interests are recognized, special strategies have to be developed to deal with the different stakeholders and to do justice to their interests. A valuable strategy to decide about these strategies is to classify the different stakeholders according their potential (high/low) to cooperate with the organization and their potential (high/low) to be a threat to the organization [13 pp. 102–105]. This analysis results in four stakeholder types and the emergence of resultant generic strategies:

**Type 1: the supportive stakeholder**
Supportive stakeholders have a high potential for cooperation and low potential for threat. The best strategy is to involve this stakeholder in the innovation of the chemical process.

**Type 2: the marginal stakeholder**
Marginal stakeholders have a low potential for both cooperation and threat. The most appropriate strategy is to monitor these stakeholders to make sure that the circumstances do not change so that they become a different stakeholder type.

**Type 3: the nonsupportive stakeholder**
Nonsupportive stakeholders are low on potential for cooperation and high on potential for threat. The recommended strategy is to defend oneself against this stakeholder.

**Type 4: the mixed-blessing stakeholder**
Mixed-blessing stakeholders are high on both, potential for cooperation and threat. This type of stakeholders could become a supportive or nonsupportive stakeholder. The preferred strategy is to collaborate with a mixed-blessing stakeholder to enhance the likelihood that they become supportive.

We would like to emphasize that the strategic importance of different stakeholders has to be analyzed during the whole innovation project, at each stage from the concept stage to the process start-up to the product launch stage, although the actual involvement of each stakeholder may vary at each stage.

Many companies use another method to characterize stakeholders. This method distinguishes stakeholders into three categories. The first category is "inform." These stakeholders have to be informed about the upcoming changes. The second category is "consult." These stakeholders have to be consulted about the ins and out of the intensified processes. In this consultation process, their interests have to be explored and added to the lists of requirements. The third category is "get agreement with." These stakeholders have to agree to the intensified process and the product that results from the intensified process. In essence, both methods have the same objectives. They try to distinguish the different types of stakeholders so as to manage them adequately.

### 4.3.3 Beliefs in society

Beliefs in society highlights the ideals and the basic beliefs of society that will influence the design of the innovation. Generally, we do not realize how important ideals and basic beliefs are. They are like air. We breathe in air every moment without realizing that it is a life supporting feedstock. The same holds for ideals and basic beliefs. They are like an "invisible hand" steering our opinions, decisions, and behavior, unconsciously.

Further, new ideals and basic beliefs can develop gradually. In the beginning, some individuals propose some ideas or express serious criticism. At a certain moment, these ideas or criticisms get momentum because they are picked up by social actions groups. Step by step, they gain momentum, grow out, and get worldwide support. Examples of ideals and basic beliefs are sustainability, circular economy, and sustainable development goals of the United Nations. The sustainable development goals was formally started in 1987 with the Brundtland report and is now a leading ideal worldwide.

One of the major expressions of the beliefs in society is the sustainable development goals of the United Nations. These goals are so important that we have presented them in Chapter 3 and we address them explicitly in every stage in Chapters 6–11.

Another expression of beliefs in society is the cultural-religious beliefs in which a (new) plant is allocated. These beliefs may vary from a country's attitude to innovation to a country's view on the use of alcohol as feedstock.

## 4.4 Analyzing existing practices and guiding future innovations

The VIB practice model has two areas of application: (a) to analyze existing industrial innovation practices and (b) use it to guide future innovation projects.

First, the VIB practice model can be used to analyze an existing industrial innovation practice. This is shown for instance in the industrial innovation cases of

Chapters 13–17. In this application this model is used to understand the state of the art. It is – to use a German expression – about the "ist" of the practice. Basically, three questions have to be asked about the past:

Which values of engineers have determined the present state of the art?
Which interests of stakeholders have dominated the industrial innovation practice?
Which beliefs in society were considered?

It has to be noted that his analysis has to be "factual." That means, it is not about wishes or intentions of the present managers or leaders, but it is organizational facts that give an answer on these three questions. The answers of these questions partly can be found in the archives, and partly in the present procedures, processes, designs and behavior.

Second, the VIB practice model also can be used to develop existing industrial innovation practices for green or sustainable process intensification innovations inside the company. It is – again a German expression – about the "soll" of the practice. We would like to stress that such a development is surrounded by all sorts of uncertainties. Despite that, it gives insights into key parameters that have to be considered. In such a situation, the aforementioned questions have to be answered in view of the future:

Which values of engineers are involved in breakthrough innovation and will be important in the future?
Which interests of stakeholders will dominate the innovation and the industrial innovation practice of the future?
Which beliefs in society have to be considered for innovation?

In Chapters 13–17, we have used this first approach in the cases. Lessons learned from that approach have been used to improve the VIB model and then in Chapters 6–11, we have provided information so that the second approach can followed for guiding future innovation projects.

## 4.5 The VIB practice model and the stage-gate innovation method

In our view, the VIB practice model can be used in every stage as outlined in Chapters 6–11. However, the focus of VIB analysis depends on the character of the stage under consideration. For example, in the discovery and concept stages, engineers can reflect on the values of engineers – Do they have the drive to pursue the concept? In the feasibility stage, when it is known where the process will be implemented, all three perspectives can be used. The governing beliefs in the local Society will be discussed in view of the consequences for the design. For instance, the choice of feedstock material from animals may change if the process is implemented in a Muslim or a Hindu dominated culture. Also, in that stage and in the later stages, an analysis of the interests of stakeholders will be carried out. By then, customers will have been identified and their needs considered. In the detailed engineering phase, emphasis has to be on translating the results of the foregoing VIB analysis into the concrete design.

# References

[1]    Ribeiro, PF, Polinder, H, and Verkerk, MJ, Planning and designing smart grids: Philosophical considerations, IEEE Technology and Society, 2012, 31(3), 34–43.
[2]    Verkerk, MJ, Ribeiro, PF, Basden, A, and Hoogland J, An explorative philosophical study of envisaging the electrical energy infrastructure of the future, Philosophia Reformata, 2018, 83, 90–110.
[3]    Van Hoof, J, Aging-in-place. The integrated design of housing facilities for people with dementia, PhD thesis, Technical University of Eindhoven, Eindhoven, 2010.
[4]    Van Hoof, J, and Verkerk, MJ, Developing an integrated design modelincorporating technology philosophy for the design of healthcare environments: A case analysis of facilities for psychogeriatric and psychiatric care in The Netherlands, Technology in Society, 2013, 35(1), 1–13. http://dx.doi.org/10.1016/j.techsoc.2012.11.002
[5]    Verkerk, MJ, van Hoof, J, Aarts, S, de Koning, SJ, and van der Plaats, JJ, A neurological and philosophical perspective on the design of environments and technology for older people with dementia, Journal of Enabling Technologies, 2018, 12(2), 57–75. https://doi.org/10.1108/JET-11-2017-0043.
[6]    Verkerk, MJ, Holtkamp, FC, Wouters, EJ, and van Hoof, J. Professional practices and user practices: An explorative study in health care, Philosophia Reformata, Dec 2017, 12; 82(2), 167–191.
[7]    Holtkamp, FC, Wouters, EJM, and Verkerk, MJ, Understanding User Practices When Drawing up Requirements – The Case of Designing Assistive Devices for Mobility, International Journal of Environmental Research and Public Health, 2019, 16, 1–12. doi:10.3390/ijerph16030318.
[8]    MacIntyre, A, After virtue. A study in moral theory. 2nd ed., University of Notre Dame Press, Notre Dame, USA, 1984.
[9]    Schatzki, TR, Social practices. A wittgensteinian approach to human activity and the social, Cambridge University Press, New York, 1996.
[10]   Nicolini, D, Practice theory, work, & organization, an introduction, Oxford University Press, Oxford, 2012
[11]   De Vries, MJ, and Jochemsen, H, (eds.), The normative nature of social practices and ethics in professional environments, IGI Publishers, Hershey, PA, 2019.
[12]   Collins, J, and Porras, JI, Built to last: Successful habits of visionary companies, Harper Business, New York, 1994.
[13]   Buchholtz, AK, and Carroll, AB, 7th ed., Business & society, South Western, Cengage learning, 2009.
[14]   Harmsen, J, De Haan, AB, and Swinkels, PLJ, Product and process design. Driving innovation, De Gruyter, Berlin, 2018.

# 5 Multidisciplinary assessment – integrating different knowledge types

## 5.1 Introduction

Let me start with a story from my own memory (Jan Harmsen) that illustrates the need for integrating different types of knowledge in innovation projects. Shell Chemicals developed a breakthrough novel-intensified process for the production of an existing bulk chemicals product, completely specified at a scale of 30 kt/year. The new process consisted of a liquid–liquid reactive extraction, multistage mechanically mixing column, that was operated continuously. A novel catalyst was used to improve the reaction rate. The existing process was a mechanically stirred batch reactor. The development, design, and construction took several years. The process construction was on its way when, in the quarterly meetings between operations (OPS) and marketing of Shell Chemicals, the marketing man said: "Should we not send a product sample to one of our main customers, just to ensure him that with the new process the product will remain the same?" All agreed, and a batch of 200 kg was made using the new catalyst in the laboratory at the Shell Pernis site using an available pilot plant batch reactor. The product met all specifications, and the batch was confidently sent to the customer.

The customer reacted violently. "This new product is unacceptable to me. Its organic chlorine content is 20 ppm, while I normally get a product with an organic chlorine content of 165 ppm with a very small variation." Shell reacted by stating that the product met the specification of <170 ppm. The client reacted even more strongly and said that his catalytic process is very sensitive to the organic chlorine (OC) content and that the variation should be within 5 ppm around the 165 ppm. The problem was then conveyed to the process development engineer. He came to me and explained that he could not meet this customer requirement, as the continuous process was fed from three batch reactors, each with a batch time of 4 h. The feed from these batch reactors varied in OC content and the measurement of that OC took several hours. Hence, controlling the new process with the input data was impossible.

I advised him to solve the problem using the Taguchi design approach. This means that the control problem is changed to a design problem, by which, the design is made robust to input disturbances. Four design process parameters – alkaline surplus, alkaline concentration, stirrer speed, and temperature where available. A set of parameter values was then found for which the new process became sufficiently robust to input variations. In addition, a statistical process control method was developed for any longer term drift in product quality. All process operators were instructed on the new method of process control. Start-up was rapid and uneventful. The client was satisfied with the product from the new process.

https://doi.org/10.1515/9783110657357-005

This story reveals a narrow escape with a lot of luck. The whole discussion between marketing, process developers, process designers, and operation should have been held in the process development stage and a product sample should have been sent at that stage. In that way, the problem about the not sufficiently specified product would have surfaced at the development stage, and adequate design and control measures would have been taken.

This is not an isolated story. There are many stories about innovation projects and stage gate decisions in which not all disciplines were represented. Also, there are stories about differences of opinions that were not defined as a source of knowledge but were seen as troublesome with the result that the minority accepts a majority position. Finally, there are stories where management overrules engineers in making decision. Often, these types of situations do not run into problems. However, at times, they do. In our view, the risks of process intensification are too high to accept these types of situations. This is especially true in cases where there are different insights with respect to the feasibility of an intensified process, a safety-health-environment assessment, or a risk assessment. For that reason, we do emphasize that different practices having different types of knowledge need dialogues are required to settle differences of opinion.

## 5.2 Different practices: R&D, operations, marketing, and sales

It is well known that R&D engineers believe that the future of the company fully depends on their ingenuity to develop new processes and new products. It is also well known that marketers and salesmen create the impression that the whole business circles around their work. Finally, factory managers often emphasize that all money is earned in the factory. Why does every function think from its own perspective? Why does every function easily overlook the specific contribution of other functions? These departments have, in the words of Nicolini [1] (2012), different "patterns of doing and saying." As a result, different subcultures develop. In other words, R&D, OPS, and marketing and sales (M&S) are different practices.

In this section, we use the VIB practice model of Chapter 4 to show that R&D, OPS, and M&S are different practices. We will show that each practice generates different kinds of knowledge that have to be integrated into the innovation process.

The objective of R&D is to develop new processes or improve existing processes, products, and services. These activities have a technological nature and require an innovative and creative attitude. The most important values of engineers here are efficiency, effectivity, safety, health, and environmental and control (removal of uncertainty). However, for the R&D department, the value "creativity" is also important. Perhaps, also the value "empathy" in the sense of understanding the needs of customers has to be mentioned. R&D has to also address actively the interests of stakeholders. Generally, R&D has to focus on stakeholders that support their innovation process, for example, knowledge institutions, manufacturers of equipment, and innovative customers. Finally, beliefs in society are also important for R&D. Especially, they have to consider the sustainable development goals as proposed by the United Nations. They have to also explore the future in-depth, as the new process is implemented several years after the R&D stages and will then be operated for a long time.

The objective of OPS is to produce existing products in a reliable and efficient way. This activity requires a "conservative" attitude: OPS is a repetitive activity that preferably runs without any changes in settings, conditions, and quality of feedstock. These activities also have a technological character. The most important values of engineers in OPS are control, efficiency, effectivity, safety, and health. OPS has to also address the interests of stakeholders actively. Especially, stakeholders such as suppliers of feedstock, customers, and companies for repair and maintenance are relevant. Two types of beliefs in society are also important for OPS. First, OPS has to cope with cultural-religious rules, habits of operators, and the local society. Second, it needs to have an open mind for new ideals or basic beliefs that are "in the air" and may result in additional regulations for their process.

The most important objective of M&S is to sell the products of the company at a profitable price for a long time. To that end, it has to investigate the future needs and wishes of customers. The nature of these activities has an economic character. Generally, the values of engineers such as control, efficiency, effectivity, safety, and health (see Section 3.5) also hold true for M&S. On the one hand, customers appreciate that the supplier controls the process that is efficient and effective, and complies with safety and health regulations. On the other hand, these values also have to be specified for the transport and delivery process. Additionally, the value "empathy" may be important to understand the wishes of the customer. M&S has to also cope with interests of stakeholders. Their most important stakeholders are the present customers and potential customers. Finally, also M&S has to fathom the beliefs in society as they come to the fore in the discussions with their (potential) customers.

Our conclusion is that R&D, OPS, and M&S departments are different practices. First of all, the nature of their activities is different. Second, because the VIB practice model shows that values of engineers are only partly overlapping in these departments. Furthermore, each department has its own set of stakeholders, and their focus with respect to beliefs in society is different.

## 5.3 Different practices: supporting departments

In the foregoing section, we have argued that the practices of R&D, OPS, and M&S are different. As a consequence, these practices "produce" different kinds of knowledge that are needed for industrial innovation.

A similar analysis can be carried out for the supporting departments such as purchasing, safety, human resources, financial control, and maintenance. First, for all these departments, it holds that the nature of their activities is different. For example, the activities of human resources department focus on training, quality, and health of the employees. It has a social and moral character. Maintenance, on the other hand, has a quite technological character. Second, a VIB analysis (Chapter 4) will show that there are remarkable differences in values of engineers (better: values of supporting departments) and in interests of stakeholders, and how they address beliefs in society.

In the context of this book, it is not necessary to do a more detailed analysis of the nature of the supporting departments. It suffices to conclude that each supporting department is a specific practice that can contribute specific knowledge to the overall industrial innovation practice.

## 5.4 Integration of knowledge

Industrial innovation practices need to integrate the different bodies of knowledge of the different practices in the company, not just the knowledge of the practices of R&D, OPS, and M&S, but also the knowledge of supporting departments.

The following three rules facilitate the integration of knowledge for industrial innovation projects:

1) *Creating room for differences.* Each department (R&D, OPS, M&S, relevant supporting departments) makes an analysis of the innovation project from its own perspective.
2) *Making differences explicit.* In a joint meeting, the different analyses are shared with each other. The specific perspective of each department is made explicit.
3) *Integrating different types of knowledge.* The specific perspectives of all departments are integrated in the project plans and in the decision-making processes. In case of disagreements or conflicting requirements, a further analysis is required.

## Reference

[1]  Nicolini, D, Practice theory, work, &organisation, an introduction,Oxford University Press, Oxford, 2012.

Part B: **Application of theory: guidelines and methods**

# 6 Discovery stage

## 6.1 Introduction

The discovery stage is about obtaining new ideas and working them out to such an extent that they are understandable to others. In general, the discovery stage does not follow a well-defined path with stepping-stones, but is meandering like a new stream generating its own river bed. In large companies with an R&D department, various structures are available to create freedom for the inventor to generate and try new ideas, with little or no management involvement. Start-up companies often have an inventor with a great idea which she pursues with tenacity. Finding a market for the idea is often one of the problems that needs to be solved. It may involve redefining the process target, making a simple design sketch to explain what the invention is about, or performing additional experiments to convince the inventor and others that the idea is sound.

A brilliant description of the thinking and working of such an inventor can be seen in the autobiography of James Dyson, the inventor of the Dyson vacuum cleaner[1]. His invention is based on using a cyclone to separate dust particles from air, rather than a filter. This makes the vacuum cleaner far more efficient, more reliable (it maintains a uniform suction power in time), smaller in size, and comes at a lower cost to the owner and to the providing company, making it a process-intensified innovation.

I read Dyson's book of 291 pages when I was on a holiday. His motivation to develop his process against all odds lay in making a beautiful thing and a little money. More and more young people have a similar drive, often with the focus on a thing that will help the world to be a better place. The cases described in Chapters 13–17 also reveal this engineering drive to make something new and useful.

This chapter consists of a description of project steps in the discovery stage in Section 6.2 and a description of supporting methods for executing these project steps in Sections 6.3–6.7. Figure 6.1 shows the connections between the project steps and the methods.

The project steps and methods in this chapter should be seen as guidelines that facilitate the inventor and her environment to get the idea to the next stage. The steps may be taken in the presented order. However, racing to the last step or trying to jump the hurdle to the next stage is okay. For most inventors, learning by trial-and-error is often the only way that suits them.

https://doi.org/10.1515/9783110657357-006

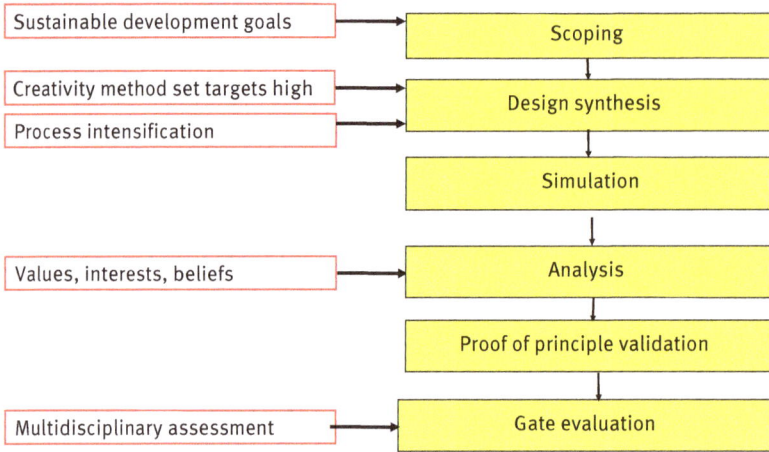

**Figure 6.1:** Discovery stage: steps and methods.

## 6.2 Project steps discovery stage

### 6.2.1 Scoping

Scoping for the discovery stage is hardly applied formally in practice. The inventor in general starts with an idea with little conscious deliberation on why he wants to work on this idea and whether the idea fits the company objectives or fits the society contexts. He may want to contribute to a better world by solving an urgent or important problem. He may be driven by an environmental concern over an existing process. He may just be excited with an idea and later may find a justification for the idea. If he works for a large company with a large R&D department, he will find a way to work on the idea within or outside the budget structure. If he works for a small company, he will discuss the idea with the company head. If he is the company head, he will find others to work on his idea.

The inventor may find it helpful to sell his idea by looking up the company business strategy, and if available, the company R&D strategy; he may then write a simple scope statement showing how his idea fits these strategies.

Further, he may look up the sustainable development goals shown in chapter 3 and select the SDGs that are relevant to his idea and then also include these in the scope.

He may also consider the design boundaries of his idea. Is it a process section within a larger existing process? Or is it an entirely new process concept?

Anyway, the advice is to write a scope statement, containing one or more of the elements discussed above. It will help the inventor and his environment in the endeavor to turn the idea into reality.

## 6.2.2 Design synthesis

Design synthesis of the process idea depends on the process scope. The scope can be an entirely new process. If that is the case, it is best to identify the most difficult part of the process, find information that helps solve that problem, and then generate several design ideas.

Ideas may be generated by holding brainstorming sessions. A variant that is often more successful is holding brainwrite sessions, where each member writes an idea. The papers are then shared with other members who enrich the ideas. After a few rounds, several good ideas are generated. The advantage of brainwriting is that it works better than brainstorming for introverts – process engineers are often introverts.

Harmsen provides several more ways of generating process ideas in the discovery stage [2].

If the scope is about a new reaction system, then considering the phases (gas, liquid, solid) and the fundamental functions involved, a reaction system may be sketched.

If the scope is about a new separation, then a sketch of the essential steps in the separation can be made.

Section 6.4 provides PI methods for generating breakthrough new process ideas and also PI methods for improving existing processes and available process designs.

## 6.2.3 Simulation

Here, the meaning of the word "simulation" is to consider the working of the idea in people's minds (and not so much about making a mathematical model and then simulation of the working with that model). Most people can do this simulation. Making sketches of the working principle can help in this type of simulation. Often, this simulation generates additional ideas, increasing the chances of success.

## 6.2.4 Analysis

The ideas generated in the synthesis step may be ranked using a few criteria or by the requirements stated in the scope. Also, the ideas may be analyzed by drawing sketches of the process blocks and stream connections between them. If the new process idea involves a separation step with a separation agent, such as a solvent or an adsorbent then also ways of recovering the separating agent should be considered. If a solvent is involved, then available streams such as feed streams, product streams, and internal process streams should be considered as alternatives to the solvent. If the separation involves stripping with nitrogen or air, then an alternative such as distillation should be considered, as stripping gases always needs a second step to recover the stripped material. The same holds for the use of water as a washing agent to remove components. Alternative washing media such as available streams should be considered.

### 6.2.5 Proof of principle validation

Ideas generated and analyzed to some extent need proof that they can work in practice. If physical or chemical interaction between molecules, or a chemical reaction, or complex hydrodynamic behavior is involved, then this proof needs to be experimental proof, as theoretical predictions of these interactions are still very poor. It is left to the ingenuity of the researcher to make a simple experimental laboratory setup to prove the working principle that will convince others, who will provide funding for further development of the principle.

If no molecular interaction or complex fluid flow behavior is involved, such as in a novel heat exchanger, then a mathematical model may provide the proof of principle. This however requires knowledgeable managers to accept this type of proof. Often, in these cases, experimental proof will also be needed.

An example from my own experience on using a model as proof of principle is the oil from shale process concept generated by Heinz Voetter in Shell. He invented a complete process starting from crude shale. The process consisted of fluidization process steps only. It involved a fluid bed heat changer, a fluid bed retorting, cracking the kerogen and releasing the oil vapor from the shale, a fluid bed riser combustor, and a fluid bed nitrogen slide for transporting spent shale from the riser combustor. Part of the hot spent shale was used to heat the fresh shale in the retort. The remainder was used to pre-heat the fresh shale. All process heat was generated by the fluid bed combustor in which the last bit of residual kerogen coke was burned.

The process was simulated with a flow sheeter program, also developed by him. Patents were granted on the simulated results as proof for the working of the process. An economic analysis showed that this process concept was far lower in cost per ton of oil produced, mainly because the commercial scale with a single train of main equipment could be designed for 80 000 t/day of crude shale, while existing single train processes of competitors were, at least, smaller in capacity by a factor of 7.

Voetter could convince top management to provide 25 M$ for further development of the process because each process step was based on fluidization, a very well-known process principle in oil refineries, specifically fluid catalytic cracking with a similar solids flow capacity as the oil shale process and because his computer simulation showed the feasibility of the heat balance.

In a way, the Shell shale process can also be seen as process-intensification, as it was much smaller in size per ton of product than all existing shale processes, due to the following steps: fast fluid bed retorting, using fluid riser reactor to combust residual carbon from the shale particles, using hot spent shale as heating medium, and heat exchange in fluid beds, with high heat transfer coefficients. Also, the yield of oil on fresh shale was higher due to the rapid release of oil vapor and rapid subsequent cooling, reducing undesired cracking reactions.

### 6.2.6 Stage-gate assessment

Stage-gate assessment of discovery projects is carried out in large companies by a panel of experienced managers. In general, the assessment can be on "go/no go" decisions of technical feasibility, health, and safety aspects. Furthermore, the project will be ranked with many other projects using economic potential, business strategy fit, and now (more and more) including Sustainable Development Goals contributions. The highest ranked projects will receive further funding for the concept stage. Section 6.6 shows an improved way of this stage-gate assessment involving a multidisciplinary team for the assessment.

## 6.3 Sustainable development goals method

The process industry is strongly dependent on a continuous stream of new ideas to feed their innovation funnel. The SDGs can function in two different ways:
1) as a stimulus to come up with new innovative ideas;
2) as a source of guidelines that can be used to investigate the full potential of a new innovative idea that was generated in a general brainstorm session or in a PI brainstorm session.

**SDGs as a stimulus to come up with new innovative ideas**
The most practical way to use SDGs as a stimulus to come up with new innovative ideas is to discuss every SDG from this perspective. The discussion can be done in two rounds. In the first round, every SDG is discussed in short. The description in Section 3.2 is a starting point. The main question is "Will SDG help us generate new ideas?" Take, for instance, goal SDG 5 (Achieve gender equality and empower all women and girls). Indirectly, this will help engineers generate new ideas for process intensification in the chemical process industry, because women on average are more careful in taking into account social aspects in design, which stimulates generation of new designs. Also, goals like SDG 12 (Ensure sustainable consumption and production patterns) and SDG 13 (Take urgent action to combat climate change

and its impacts) challenge engineers to develop sustainable production patterns and to combat climate change by means of process intensification. The first round will result in a number that SDGs whose potential has to be discussed in detail. In the second round, every selected SDG is discussed in more detail. How does this SDG help generate new innovative ideas? In this round, it may be very worthwhile to discuss the detailed elaboration of the SDG concerned as described in *The 2030 agenda*. The result of this round has to be a number of innovative ideas for closer consideration.

**SDGs as a source of guidelines to investigate the full potential of new innovative ideas**

In this case, one of the brainstorming sessions results in an innovative good idea for process intensification. Then, the full potential of this innovative idea is investigated by analyzing this idea from the perspective of SDGs. This discussion too, can be done in two rounds. In the first round, the potential of every SDG is discussed to refine and enrich the new innovative idea. In the second round, all selected SDGs are discussed in more detail. The result of this round will be an innovative idea that is refined by and enriched with sustainable development considerations.

## 6.4 PI design synthesis discovery method

### 6.4.1 Breakthrough PI design synthesis method

Breakthrough process designs can start in the discovery stage using PI design ideas in an intuitive way, by just reading parts of Chapter 2. It can also be done in a systematic way. The latter follows a logical sequence:
1) Plant-wide spatial
2) Functional synergy
3) Temporal
4) Thermodynamic
5) Spatial equipment

The logic behind this sequence follows. The plant-wide spatial domain views the whole design opportunity space and generates options that affect more than one process step. In the plant-wide spatial domain, one looks for opportunities to minimize by-product formation and the number of process steps. This can be done by exploring novel reaction systems producing the desired product only. In this way, the lowest cost on feedstock and the lowest number of separations are obtained. This is the plant-wide process intensification approach as described in Chapter 2. The new reaction system may be based on a catalytic conversion, an enzymatic conversion or micro-organism conversion. The advantages of such a process are that it has the lowest raw materials cost per ton of product and that it requires less separation steps. Chapter 2 provides the plant-wide PI options. Chapter 15 shows an example of what such a plant-wide process-intensification design can mean for savings in feedstock cost, investment cost, waste reduction, and energy requirements.

When the reaction system has been explored in combination with the separation options, then the next PI domain to be considered is functional synergy. This means that all essential process functions such as reaction function, separation function, mass movement function, and product formation functions are placed in a block-flow diagram and connected to each other by streams, so that finally a complete block-flow process diagram is generated. The combining functions in a single confinement are explored such as in a reactive distillation column. In particular, the use of streams for extraction or stripping purposes should be explored. Chapter 2 provides the theory of function integration. Chapter 14 shows a powerful example of what can be achieved if all streams are optimally used, so that the whole process can be carried out in a single column, as opposed to 11 separate process units.

Then, the temporal domain is explored for further investment cost and energy cost reduction. Chapter 2 shows several temporal domain techniques. Examples of commercially applied techniques are reverse flow combustion and cyclic distillation. The latter is described in detail in Chapter 16.

If energy has to be supplied, alternative sources of energy as described for that PI domain can be explored, such as those shown in Chapter 13. Chapter 2 provides a list of options. The most promising can be selected and a proof of principle experiment can be setup and executed.

When one or more options are appealing, then the researcher can test them experimentally to see if they could work. These are called "proof-of-principle experiments." If one fails, then the next option can be tested, until at least one promising option passes the proof-of-principle test.

For this option, a process sketch with input and output streams should be made so that the option can be communicated to others and can be analyzed using the VIB perspectives of Chapter 4 and evaluated in the multidisciplinary assessment. For the discovery stage, it is attractive to have more than one option. When, in the concept stage, the option fails, another option can be picked from the discovery stage. Even if the option does not fail, another option may also be explored in the concept stage, so that a comparison can be made, and the best option chosen.

Finally, the spatial domain for equipment can be explored. Chapter 2 provides available techniques.

### 6.4.2 PI design synthesis for revamp

Revamping an existing process often starts at the process engineering department of a company, rather than in the R&D department. Often, the scope is provided by the business to increase the process capacity or the process efficiency, or both, of the existing process and using proven unit operations to obtain the design result, without requiring research or development. The subject is treated here in the discovery stage, because it is advised to challenge the provided scope by considering process-intensified options to make a bigger improvement than that envisaged, using conventional units operations.

To that end, PI options described in Chapter 2 should be considered. The approach starts from an existing design, finds the limitations of the existing design using critical performance phenomena, and then uses PI domain knowledge to overcome these limitations. This method has been derived from Stankiewicz et al. [3]. It can also be used for education purposes.

## 6.5 VIB perspectives methods

The application of the values of engineers, interests of stakeholders and beliefs of society (VIB) method, as described in Chapter 4, enriches the discovery stage. The main reason is that engineers are invited to think over new questions, especially, to think over these questions from the beginning of the discovery stage.

### 6.5.1 Values of engineers

The discovery stage is often an uncontrollable process. A few ideas come up and some of them die, but a few ideas crystallize and attain a certain form. At that moment, the inventor does all kinds of mind experiments to elaborate and to refine the original at idea. Still, the thinking space of the engineer is very large.

The reflection on values stimulates the engineer continuously to think over the "why" of an innovation. Why do I want to realize a better process? The case studies in this book shows that there are "hidden" motives and "hidden" values. For example, in the case study on Eastman 14), a hidden motive was to be so innovative as to generate jobs for the local region. Another example is the case study of cyclic distillation that leads to ethanol with less (toxic) contaminants for human consumption (chapter 16). It is very important to discuss motives in the discovery stage. May be, these motives reveal hidden "values" that have to be made explicit in the course of the discovery stage and the stages that follow.

Section 4.5 proposes a method to reflect upon values. In addition, a general list of engineering values is already provided. It is important to note that this method has to be applied in the spirit of the discovery stage – that means, very open minded, with a lot a creativity, and in an unstructured fashion.

### 6.5.2 Interests of stakeholders

In the discovery stage, the interests of the most important stakeholders have to be considered. There are two reasons. The first is that it invites the engineers already in the first phase to think from the perspective of the different stakeholders. The most challenging question is: How to add more value to the most important stakeholders? The second reason is that it supports engineers to sell their ideas to higher management and to increase the quality of the multidisciplinary assessment at the end of the discovery phase.

Section 4.6 proposes a method to reflect upon the interests of stakeholders. In this stage, the most important questions are:
1) Who are the most relevant stakeholders for the innovative idea?
2) How to increase the value addition for these stakeholders?

The reflection on these questions has to be done in agreement with the spirit of the discovery stage. That means, intuition, creativity, and empathy are the most important ingredients.

### 6.5.3 Beliefs in society

Beliefs in society refers to the cultural and religious beliefs of the local society in which the plant with the process-intensified process will be allocated. In this stage, it is not necessary to think over a possible allocation. The focus is on the fruitfullness and the business opportunities of the new idea. For that reason, beliefs in society may be discussed in this stage, although, if the final destination of the process is known, it could be considered.

## 6.6 Multidisciplinary assessment method

In Chapter 5 of this book, we have provided the multidisciplinary assessment method. The key aspect of the method is to integrate the different bodies of knowledge from the beginning in the innovation process – especially the knowledge of the practices of R&D, OPS, and marketing & sales (M&S) can be applied in the discovery stage-gate, to a certain extent.

Stage-gates assessment differ enormously across companies. Large companies, in general, have an annual assessment of all discovery stage research projects by a panel, in which they decide which to stop, which to continue, and which to start. In most cases, managers from R&D and from business departments are part of the panel. In medium-sized companies too, OPS managers are, often, part of the panel. In small companies, often, there is no stage-gate. An idea is pursued till it becomes clear to the director of the company that it will not work, or he starts to believe in the project, becomes heavily involved, and pursues it against all odds to make it a success. Chapter 13 describes such a case.

In this section, we provide a multidisciplinary assessment for any company size, although we have used the language and organization of a medium to large company with R&D, OPS and manufacturing & sales (M&S) departments. For small companies, the knowledge of these departments rests in a few persons.

A multidisciplinary assessment for this stage-gate can be done in the following way:

1) The inventor discusses the innovative idea with his colleagues in the R&D department. In this meeting, the technical feasibility and the development risks are discussed, mainly.

2) The inventor discusses the innovative idea with his colleagues in the OPS department. In this meeting, the conditions for technical control and production are discussed.

3) The inventor discusses the innovative idea with his colleagues in the M&S department. In this meeting, the market opportunities for the intensified process are discussed, in more detail. Particularly, the competitive edge of the product produced by process integration has to be discussed.

4) Finally, the inventor invites one or two colleagues each from the R&D, OPS, and M&S departments to integrate the results of all earlier discussions in a multidisciplinary assessment report.

## 6.7 Creativity method: set targets high

A promising method of generating new ideas is to do a brainstorming session, in which the targets for a new solution are set high. This method is useful for the discovery stage and for the concept stage. It has been proven to be successful in the case described at the end of this section. It can be done as follows:

First, we define the parameters for which we would like to define targets. These can be derived from the Sustainable Development Goals. Often, parameters derived from the environmental SDGs such as reducing emissions to the environment, or economic SDGs such as product purity (for market growth), or social SDGs such as increased personal safety are selected.

Second, we set targets for these parameters. Generally, engineers do not like to set the targets too high. In their view, they will be judged on the final result, so generally, they think it is better to surpass the expectations than to fail in meeting the expectations. However, a danger is that "too low expectations" will not challenge the engineers enough and they will then come up with moderate improvements over the existing process. Those moderate improvements are then not sufficient to pay for the development cost, and the whole project will quickly be terminated. For this reason, high targets are needed to obtain breakthrough design. In my previous company, we often set targets for new process concepts by stating that the new concept should be lower in cost and emissions by a factor of 4. In fact, the targets must be set so high that everybody says: that's impossible or that's nearly impossible to achieve.

Third, we start with brainstorming about PI and PI principles for the relevant process.

The main idea about "set the targets high!" is to set the targets so high that everybody is aware that it is useless to suggest a rather "simple" adaptation of the process or using conventional unit operations based design methods. On the contrary, everybody is invited to think in terms of breakthroughs. A reduction of a process from 10 to nine steps (in which only two processes are intensified) is a good improvement, but not a breakthrough. A breakthrough requires a reduction of ten to four steps or something similar to that.

In the case study, described below in Section 6.8, one of the authors (Maarten Verkerk) was the project manager responsible for the design and the building of the new factory, where this high target setting was applied.

## 6.8 Case study: breakthrough in the process "casting the foil"

In the summer of 2000, Philips sold some factories in which *multilayer ceramic capacitors* (MLCCs) were made to the Taiwanese firm, Yageo. In the light of the strong growth in the market for electronic components, the new owner decided on a substantial expansion of the capacity of the establishment in the Netherlands. To this end, a new factory had to be built. The production would be increased in one year from the existing capacity of twelve billion to forty-eight billion units per annum, so a factor 4 higher capacity for the new plant.

In this case study, we take a closer look at considerations that played a role in the process of designing the new factory. Various stakeholders were involved in the design of a factory: the employees, the clients, the parent company, the municipality, and the province as also the immediate environment (neighborhood). It was a complex design problem during which an attempt was made to synchronize social, economic, and technological considerations. To give an insight into this process, we start with a short description of the then-existing manufacturing process. Subsequently, we focus on the requirements that were set on the new process design "casting the foil."

An MLCC is a modern type of capacitor product, composed of a great number of very thin layers of alternating conductive and insulating material. The conductive materials are metals with a thickness of about one micrometer, while the insulating materials consist of a mixture of oxides with a thickness of ten to eighty micrometers. The smallest MLCC is hardly one millimeter in size, and the largest is about five millimeters.

The existing production process had five steps, every step in turn is made up of several sub-steps

1) The first step was *casting the foil*. Diverse materials were added to ceramic powder (an oxide). This was ground until the particles have a specific size. The whole was then mixed with an organic binder (a kind of liquid plastic) and some chemicals until a white liquid slib emerged. This slib was then "spread" very thinly over a metal band that went through a tunnel oven. The moisture evaporated and the result was a very thin layer: the ceramic foil. This thin white foil was wound around a spool.

2) The second step was the *screen-printing, pressing and cutting*. During screen-printing the ceramic foil was cut up into sheets, screen-printed (application of the conductive materials) and stacked on top of one another. Under high pressure, these loose stacks were pressed into small thin plates of ten by ten centimeters. These were then cut into loose products in the cutting machine.

3) The third step was to get rid of the binding material by heating and baking. First, the organic binding material added during the mixing in the first step was removed by a thermic process. This was done by heating the product slowly to a high temperature, so that the binding material was "combusted" in a controlled manner. Then, came baking. During this process, the ceramic powder particles were baked onto one another at a high temperature. The products now obtained the desired mechanical and electrical properties.

4) The fourth step was applying the electrical contacts, baking and galvanizing. Each product was provided with two contacts, one on each side of the capacitor. These contacts were needed to connect the electrodes with one another and to enable the product to contact the electric circuit in which it will later be placed.

5) The last step was *measuring and packing*. Electric product properties were measured, and some quality measurements done. Then the products were packed using a tape.

What would the new factory have to look like? A copy of the old one, only each process equipment four times large in size? Or would a new concept be needed? In the initial phase of the project, these questions were discussed in detail, for it is known from the literature that the design of a factory has a great influence on its final performance. Here, we were not concerned only with technological performance such as quality and yield, but also with social "performance" such as satisfaction and operator absenteeism. Besides, the immediate environment (the factory stood in a residential neighborhood) and the natural environment (removal of heavy metals and the emission of gases) had to be considered.

The project manager discussed the design of the process "casting the foil" with the engineers. He gave them the following targets:

1) Increase the process volume by a factor of 2, at a capacity increase of a factor of 4, so use an intensification of a factor of 2.
2) The emission of waste gases per ton of product must be reduced by a factor of 4 so that the total amount is the same as for the old factory.
3) The emission of heavy metals via wastewater per ton of product must be reduced by a factor of 4 so that the total amount is the same as for the old factory.
4) The operator efficiency must be increased by a factor of 2–3.
5) From an ergonomic point of view, all manual lifting of heavy bags must be removed.
6) The probability of operating mistakes must be reduced by a factor of 5.

The first reaction of the engineers was, "these targets are too high. It is impossible." In the end, and reluctantly, they accepted the targets and started to brainstorm. And, after a couple of weeks they came up with the first concept design that roughly met the targets. About a year later, the new detailed design still appeared to fit the target sets. As a conclusion, setting the targets high induced the mindset to think of breakthroughs [4].

# References

[1] Dyson, J, Against all odds – An autobiography, Orion Business Books, London, 1997.
[2] Harmsen, J, De Haan, AB, and Swinkels, PLJ, Product and process design, Driving Innovation, De Gruyter, Berlin, 2018.
[3] Stankiewicz, A, Van Gerven, T, and Stefanidis, G, The fundamentals of process intensification, John Wiley & Sons, Hoboken, 2019.
[4] Verkerk, MJ, et al., Philosophy of technology. An Introduction for technology and business students, Routledge, London, 2016, 185–191.

# 7 Concept stage

## 7.1 Introduction

The main purpose of the concept stage is to provide a process design containing all essential technologies, with all the critical aspects for commercial-scale implementation so that a multidisciplinary management group can decide to continue the project or terminate it.

The project steps to be taken are scoping, design, modeling, analysis, experimental validation, stage-gate evaluation. These steps are described in Section 7.2 and form the project backbone. Innovation methods forming the project body are presented in Sections 7.3–7.8. Figure 7.1 shows how the steps and methods are connected.

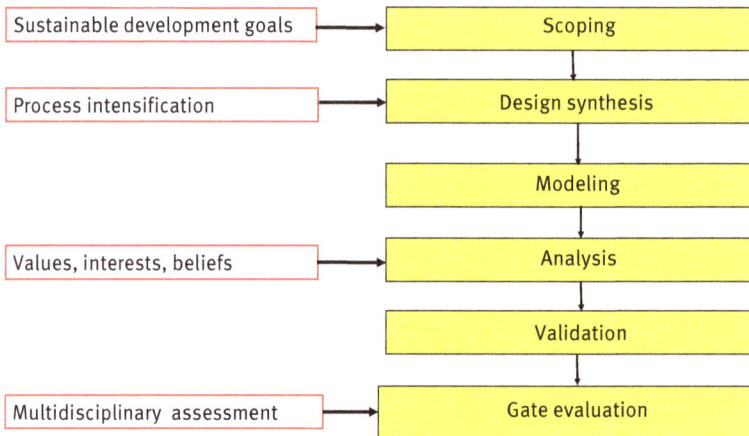

**Figure 7.1:** Concept stage: steps and methods.

## 7.2 Project steps

### 7.2.1 Project steps background

The project steps to be taken in the concept stage are shown in Figure 7.1. These steps are derived from the process design textbooks [1, 2]. For projects involving breakthrough process intensification, the design synthesis step and the modeling step play a far more important role than in conventional process designs. Hence, they are described in detail in the subsequent sections. Also, the analysis step plays

https://doi.org/10.1515/9783110657357-007

a prominent role in the process concept design and is therefore also described in detail. Experimental validation of the process concept and its model are also dealt in detail in the next sections.

## 7.2.2 Scoping

The scoping step involves setting the goals and constraints of the project concept stage. The main purpose of the concept stage is to generate the best process concept and to find out if the process concept is technically sound and can meet all requirements. The requirements depend on the specific company and its context. The context has the following dimensions, at the minimum: safety, health, environment, economics, and sustainable development goals. For each dimension, the requirements have to be defined. Some companies indicate what should be included in the economics while others leave it open. In general, the accuracy of data required to meet each requirement will be low in the concept stage.

Data is thus gathered to allow an assessment at the stage-gate whether the requirements are all met indeed. For laying the contribution to the sustainable development goals, a dedicated method is provided in Section 7.3.

For a breakthrough process innovation, requirements should be tight. It is enormously helpful if a reference process is identified against which the breakthrough process innovation should be measured upon for all mentioned requirements. The reference case can be the best commercial-scale process in operation. This can be an in-house process or a competitor process. It may be even better to use an improved version of the reference process as the reference case, as it is likely that competitors are also improving the existing process.

The scoping also involves defining the product specifications. If the product is already in the market, then specifications can be obtained from the information available in the marketing department. It will also be available from competitor websites. For defining product specifications for a design course, finding these specifications on the internet can be a revelation for students. They will understand how many items make the specification list. For a concept design, they will then have to decide which specifications they can meet and for which specifications they will be unable to meet due to lack of information generated from their process design.

Scoping also involves defining the feed materials to the process. If they cannot yet be defined, set constraints to the feed materials that are to be selected. In the design synthesis step, various feed material to be used in design options can then be considered, analyzed, and the best feed material in combination with the design can then be selected. This feed plus design selection will most likely also involve sustainable development considerations for social, ecological, and economic impacts on society and environment. A method is provided in Section 7.3.

Scoping for revamp projects which means that an existing process has to be modified contains the same elements as for a project started from scratch. Hence, it is not only a brown-field project but has even more restrictions as it has to fit into an existing process. Therefore, the number of constraints is far larger and more diverse. Plot size limitations, for instance, are often very important. Process intensification solutions then have the additional advantage of requiring less space.

Scoping for modifying an existing process design rather then for a new to be built process at a green or brown site is also often a desired option. The perceived advantage is that the risks of implementation failure are lower than for a breakthrough novel design. This risk reduction is, however, only valid if all necessary precautions for all process innovation steps are taken. When these precautions are not taken, the modified process will still fail. Examples of the latter are provided by Harmsen [3].

Students find it a challenge to define scope, while learning. It takes some effort by the teacher to explain the importance of setting high targets. It takes even more effort to convince students to set high targets for their own course project as they are afraid that by not meeting these targets at the end of the project, they will get a lower mark. But if they set high targets, after some time, they will get very good ideas and see the advantage of having set these high targets. Hence, patience with students is also needed. The same holds for training young engineers inside companies.

### 7.2.3 Design synthesis

Design synthesis is similar to chemicals synthesis in that, it is unknown, at the start, how the process will be designed. In this design synthesis step, process options are generated, modified, and selected, until at least one design meets the requirements. The number of different process design options that can be generated using conventional unit operations as building blocks is already between ten thousand and a million [4]. With process-intensified ways of designing, this number increases enormously. To handle this problem, two approaches are available and the most suitable is selected.

#### 7.2.3.1 Hierarchy process synthesis method
The hierarchy method is based on the general reasoning that it is best to start designing from what is already known and fixed. From here, generate in sequential steps, the preliminary design options and after most preliminary steps have been taken, review the result. Modify and refine the steps until one or a few process options are complete solutions connecting input streams to output streams.

Heuristics derived by experienced process designers and based on economics help to generate the process step options and provide criteria for selecting the best intermediate options. This method is described by Sieder [1], Harmsen [2], and Douglas [4]. This method can also be supported by artificial intelligence computer programs such as PROSYN, providing hints to the designer for each process step decision in the design.

### 7.2.3.2 Super structure process synthesis method
This method is entirely based on computer programs. First, a super structure is generated, containing all process units conceivable with all potential connections. Second, by applying mixed-integer nonlinear programming (MINLP), the economic optimum unit operations are calculated.

A variant of this is provided by Demirel, considering process intensification. He first defines fundamental building blocks in the process design. After that, the designer can decide to select for each building block, a unit operation or uses process intensification ideas to fill the building blocks and combine building blocks (function synergy). Demirel then uses MINLP to optimize the process design, be it based on conventional unit operations or on process-intensified options [5].

### 7.2.3.3 Selected synthesis method for process concept design
For process-intensified synthesis, both concept design methods are not directly applicable as PI is not based on conventional unit operations. Even more important, not only are economic criteria points of concern in process synthesis but also are many others, as set in the project scope. However, one basic idea of the hierarchy method that is applicable is the idea of fundamental building blocks by Demirel [5] for process synthesis. It is worked out as follows.

### Step 1: Draw the boundaries of the process
It is of enormous benefit to the designer and all other contributors to first define the boundaries of the process. These boundaries are around all process steps and their internal streams including the recycle streams to be designed. These boundaries are only crossed by input streams and output streams. Figure 7.1 shows an example of drawing the process boundary box. It is a white box. Later, in the design process, the box will be filled with process elements.

### Step 2: Draw output streams from the process boundary shape and specify the product stream
The product specification can be obtained from existing product specifications available inside the company or from other manufacturers. Most websites provide the product specifications. If the information is not available because the product is novel, assume a high-purity of 99.9% or higher so that, for the time being, no significant byproducts can leave the process via the product stream.

It is also handy to set the size of the product output stream, the process capacity. If that capacity is not available for the concept stage, choose a size typical for the process branch. That size can be selected from ballpark figures of Table 7.1. For some companies, the fine chemicals size stops at 1 kton/year.

**Table 7.1:** Typical production sizes of a process for some industry branches.

| Process branch | Process capacity (product output) kton/year |
|---|---:|
| Oil refining | 1,000 |
| Bulk chemical | 100–1,000 |
| Fine chemical | 1–10 |

**Step 3: Specify the feed material and its source**

The feed may have been defined, to a certain extent, in the discovery stage. In the concept steps, however, all feeds are to be defined and also a start has to be made in selecting the sources of the feeds. Environmental concerns over the entire life cycle and circular economy thinking needs to be considered in this respect.

The required feed streams follow from the defined output stream. If the molecular composition in the output stream is different from the available input streams, a reaction step is needed. This reaction may have been selected and explored, to a certain extent, in the discovery stage. For the time being, just a stoichiometric reaction equation can be used to define the reaction function from which the input stream can be defined preliminary. Later in the project, the input stream sizing will be finalized when the complete process is designed. Subsequently, more elaborate kinetic models will be developed. Also, the sources of the feed streams will then be selected in view of sustainable design goals and striving for circular economy solutions. This is described in detail in the SDG goals methods for the feasibility stage. Some elements of that method may also be considered in the concept stage.

**Step 4: Define all other output and input streams**

The stoichiometric equation may show that a second output stream has to be identified for the byproduct. If the process requires a separation to obtain the product stream, a second output stream is inevitable. An atomic balance calculation on output streams and input streams is a simple check whether all output streams and input streams have been defined so far.

## Step 5: Define all fundamental functions to transform the input to the output streams

A very useful aid to support steps 1 to 4 is to make a figure as shown in Figure 7.2 Start with drawing a big box indicating the boundaries of the process concept. All input streams entering the process box are placed on the left side and all output streams leaving the box are placed on the right side. The figure is then easily read, left to right. Inside the box, all fundamental functions are placed. Finally, all streams connecting the inputs via the essential function blocks to the output streams are drawn. Essential function blocks only state, in a very simple way, what needs to be done by the function. For processes, the fundamental function options are obtained from Harmsen [2]:

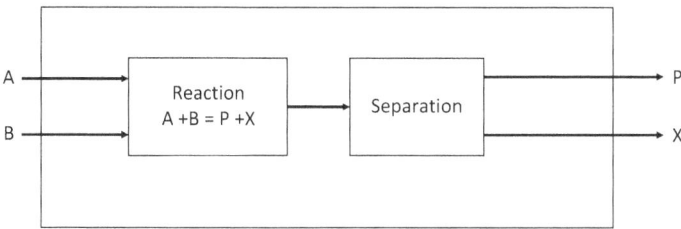

**Figure 7.2:** Example functional process Block flow scheme.

- Mass movement
- Mixing
- Heat exchange
- Reaction (molecular change)
- Mass exchange across an interface
- Separation
- Forming (shaping)

A mass balance check can be made by comparing total output with total input. In addition, an atom balance can be made for components present in small amounts in the feed stream. If, for instance, biomass is fed as input, then atom balances on sulfur and potassium can be made to see whether the needed output streams for these atoms are defined.

## Step 6: Draw recycle streams

If recycle streams are present, for instance, because a solvent is used for the reaction, and needs to be recovered and recycled back to the reaction step, or when the reaction is an equilibrium reaction so that single-pass full conversion is not feasible, then a separating function block is drawn and a recycle stream from the separation step to the reaction step is drawn as shown in Figure 7.3.

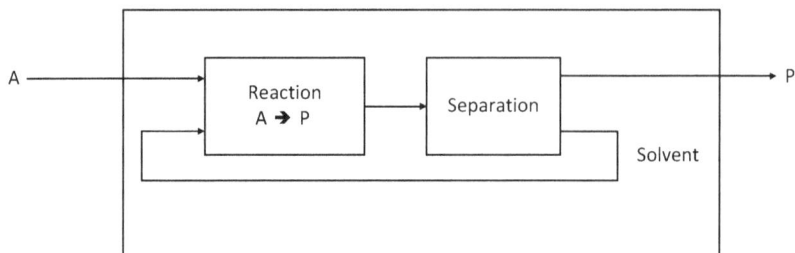

**Figure 7.3:** Functional process block flow scheme with solvent recycle stream.

For education purposes, it is of enormous benefit that the recycle stream is shown to be inside the overall process block. In this way it is clear that the input and output streams of the overall process are not affected by recycle streams. Many times, I have seen students proposing to get rid of an unwanted byproduct by having a recycle stream. By showing that recycle streams do not affect inputs and outputs, it becomes clear to them that recycling cannot remove an unwanted byproduct. For easy reading, the recycle stream back to the inlet of the reaction should be "empty" and hence, it should not contain a function block. All function blocks should be in a sequence from left to right.

An example is shown in Figure 7.3.

### Step 6: Apply process-intensified synthesis method

The process design synthesis is then completed with the process intensification concept design method described in Section 7.4, in which the essential functions are replaced in steps with real functions and in the end, with real defined equipment.

### Example case: Algae to transport fuels and use of process box for inputs and outputs

An oil and gas company desired to convert algae into transport fuels. To that end, it had called for a brainstorm meeting with marketing and technical employees to generate process plus marketing options for later R&D projects. For each idea generated, I proposed to draw process boundary boxes with product streams and algae as input stream. Everybody liked this approach. One idea was to convert lipids of the algae to transport fuels and to sell the remaining protein rich material as animal feed. We set the desired transport fuel output to 1,000 kt/year, typical for a small refinery. We then calculated the animal feed produced, from a crude estimate of the algae composition. This appeared to be of the same size. Then the marketing people realized that they needed a very large marketing organization to sell this amount of potential animal feed. They also realized that the top management would probably not fancy this idea. So, the idea of producing animal feed was dropped. Ideas to convert protein also to transport fuels were generated.

This case shows that bringing the engineering and the marketing people together in the discovery stage avoids walking into blind alleys and helps to focus the project on ideas that fit to the company's capability and market.

### 7.2.4 Modeling and simulation

Since the 1980s, process modeling has spectacularly taken off with flow sheet packages which could be bought from-the-shelf. For most conventional process unit operations, pre-programmed blocks were available so that generating a complete program for the whole process connecting all unit operations can be done in a few hours to a few days.

The obtained flow sheet model can then be used to simulate the process and find optimal conditions for each process unit operations and for the process as a whole. The model also generates compositions for each process stream. Checks on mass balances can also be performed.

For process-intensified designs, modeling, in general, is more complicated as generic pre-programmed blocks are not available. Dedicated modeling efforts are required to obtain the model. For process-intensified designs, this modeling is, in most cases, essential to determine the feasibility of the design options generated. This is particularly so for design in the temporal PI domain, as thinking dynamically about a process step is, in general, very hard.

Tian provides a nice overview of modeling and simulation for process intensification purposes with a focus of function synergy solutions such as reactive separations [6].

### 7.2.5 Analysis

Preliminary designs are analyzed first by checking if all inputs are connected via process steps to the output streams. Second, overall mass balances are checked between input and output. Third, atom balances are checked. The latter is particularly useful for trace components present in crude feed streams with inorganic components such as potassium and sulfur in biomass feed streams. Fourth, the preliminary designs are compared with the requirements stated in the scope.

Design optimization, often, with the aid of process models is also part of the analysis. The optimization targets can be by-product reduction and energy requirements. More importantly, the process design can be made more efficient by applying process intensification methods described in Section 7.4.

### 7.2.6 Validation

The finalized process concept design needs to be validated experimentally to check whether the process design does what it is intended to. This validation should be carried out with an experimental set up containing the complete process with all process units and all recycle streams. To keep the cost and the time required for this validation small, this experimental set-up will be small. If the process involves only gas and liquid flows, it will be a bench-scale laboratory set-up. Often, the process will be in glassware so that process streams can also be observed visually. Any frothing, foaming, fouling, and clogging will then be quickly identified. So, the validation only concerns flows and compositions and not yet construction material reliability.

The process model should also be validated. This is done by comparing the experimentally obtained stream compositions with model predictions. Important streams for this validation are output streams and recycle streams. Also, the output streams of main process units such as reactors and separators should be compared with model predictions.

If the model predictions deviate significantly from experimental results, the cause of the deviation should be investigated. Fitting the model parameters to the experiments should not be done. If fitting is done, the fitted model will be of little value for scale-up design as it is unreliable. Often, the fitted parameter value changes with scale.

To remember the roles of modeling and experimental validation, I quote Tony Kiss: "Models direct; Experiments decide."

### 7.2.7 Stage-gate evaluation

The purpose of the stage-gate evaluation of the concept stage is to decide whether to stop the further development of the new process or to continue its development into the next stage. A justified early-stop of a process development, at the end of the concept stage, can save the company large sums of money and reputation. A variant to stopping the project can be that the concept appears to be sound but does not fit to the company strategy [7]. In that case, the intellectual property may be licensed or sold to another company or institute.

The gate evaluation is, therefore, very important and should be carried out by a multidisciplinary panel with experienced members from, at least, marketing, OPS, and R&D departments. Business unit managers may also be present. Section 7.6 provides a method for forming this panel and for the stage-gate evaluation, considering all relevant aspects of the innovation project.

The panel will need a written report containing all results of the concept stage along with a comparison with the requirements set in the project scope. If the concept stage scope requirements could not be met, the report should recommend stopping the project. The report should also provide a list of remaining risks – unknowns and uncertainties to be addressed in the next innovation stages.

The project report will, therefore, at least treat all items mentioned in the project scope. These, probably, will include requirements to:

Fit to company's innovation strategy

Fit to company's portfolio criteria

Fit to technical reliability criteria

Fit to values of company engineers (conviction that they can develop the process)

Fit to value to customers

Fit with feedstock from suppliers and the entire value chain in view of circular economy

Fit to interest of external stakeholders

Fit to society beliefs (or a statement that this should be investigated in next stage)

Fit to safety, health, and environmental criteria

Fit to sustainable development goals (see also Section 6.3)

Fit to economic criteria

The panel will also need an oral explanation of the project's results and also needs to be able to ask questions to the development project manager. If it is a large project, then other project developers will support her in the question and answer session.

## 7.3 Sustainable development goals method

The purpose of sustainable development goals method described here is to select those that are suitable for the innovation project at hand and go a little deeper into the selected SDGs so that they become meaningful for the concept stage. Chapter 3 provides a description of the SDGs. Here, we provide some guidelines to selecting some of the SDGs and working them out for scoping the concept stage.

The concept stage starts with the end result of the discovery stage, an elaborated innovative idea. It should be remembered that the scoping requirements of the concept stage have to be considered in more detail than in the discovery stage. The main reason is that chemical installations are written off in twenty-five to fifty years. In this period, the requirements with respect to energy use, purity, waste, pollution and other parameters will become increasingly tough. As far as possible, the new design has to be "future proof."

The future proof scope can be obtained in two rounds. In the first round, every SDG is discussed in brief. The description in Section 3.2 is taken as a starting point. The main question is, does this SDG offer essential requirements for the concept phase? Goals like SDG 12 (Ensure sustainable consumption and production patterns) and SDG 13 (Take urgent action to combat climate change and its impacts) are very important to define the targets for the concept phase. They challenge engineers to think about the whole chain and set targets for the circularity of the materials flow, the amount of waste materials, the purity of the material, the energy consumption, the $CO_2$ production, and so on. The first round will result in a number of SDGs that have the potential to contribute to the targets set for the intensified process.

In the second round, every selected SDG is discussed in more detail. The detailed descriptions of the SDG as given in *The 2030 agenda* will help the engineers to identify the most important parameters and to set the targets. The result of the second round is a list of targets for the intensified process that are based on the SDGs.

## 7.4 Process intensification design method

### 7.4.1 Process intensification break-through design method

The breakthrough process intensification design starts from the essential functions block flow diagram of Section 7.2.3. The sequence of process intensification design then follows the steps described in Chapter 2. If one or more reactions function blocks are involved, then first options for those transforming these theoretical functions into real functions are considered. The option with the lowest by-product formation is then chosen.

### 7.4.2 Process intensification revamp existing process method

In many cases, a designer is asked to modify an existing process. This is known as revamping or debottlenecking. To generate a process concept, the designer can study the full scope and then make a design that fits to the constraints using theoretical inputs from Chapter 2.

Alternatively, the designer can take one promising process intensification option, work it out to some extent, compare it with some of the scope requirements, modify the design so that it meets these scope requirements, compare it with all scope requirements, and make final adjustments. In this trial and error approach, he may consult plant operators to obtain feedback. Often, this feedback also means that the scope needs further definition.

### 7.4.3 Process intensification design synthesis to modify an available process design

In many other cases, the process designer is asked to modify an available process design. The modification is often desired for an optimized process design. The difference with a revamp is that the whole process is to be constructed and implemented.

Van Gerven provides an approach for modifying existing process designs. His whole book is, in fact, based on the assumption that there is an existing design that has to be changed by process intensification methods [8]. In Chapter 8, he discusses a process intensification design teaching method for MSc students. Inspired by his method, we present here a systematic stepwise approach.

Step 1:Determine critical phenomena limitations by using the following list of potential phenomena, critical to the success of the process. The list is obtained from Harmsen [4].

**Process phenomena types critical to success**
- Feed distribution
- Residence time distribution
- Mixing
- Shear rate distribution
- Mass transfer
- Heat transfer
- Impulse transfer (pressure drop)
- Chemical reaction

This list needs little explaining except for residence time distribution. Stankiewicz states in his book that plug flow (he calls this that every molecule has the same history) is the desired optimum. However, for autocatalytic chemistry and biotechnology processes, where highest microorganism production per unit volume is desired, this is not true. Some degree of back mixing is beneficial for a defined residence time distribution. For reaction engineering, this list is known. But, it also holds for others such as separations engineering.

Step 2: Find potential process-intensified solutions to these limitations. Use the five PI domains of Chapter 2.
Step 3:Rank the potential solutions using the scoping constraints as selection criteria.
Step 4:Select the top potential solutions.
Step 5:Make a process design based on these potential solutions.

## 7.5 VIB perspectives method

The application of the values of engineers, interests of stakeholders and beliefs of society (VIB) method, as described in Chapter 3, supports the analysis step in this stage. The main reason is that engineers are invited to reflect and discuss their process proposal and their model from the different perspectives of the VIB model. In fact, engineers are invited to discuss the process proposal and model it from the perspective of the engineers, stakeholders, and beliefs in the society.

### 7.5.1 Values of engineers

The researchers, for simplicity here always called the engineers, are urged to answer the question: What values are important for me? Are these values crystallized in this worked out idea? It should be noted that in comparison with the foregoing steps, new values pop up. The reason is that in the scope step, SGDs are discussed and from there, a process design is made. While more personal questions are addressed here, it should be noted that in this phase, it is not necessary to have a shared opinion of these values. Differences of opinion, however, have to be made explicit. Generally, in the course of the project, the differences of opinion will disappear.

In Section 3.5., a method is proposed to reflect upon values. In addition, a general list of engineering values is also provided. Together with the results of the SDG method, a more complete and more focused list of engineering values will evolve.

### 7.5.2 Interests of stakeholders

In the concept stage, the interests of the stakeholders that are already known have to be considered, especially of suppliers, customers, and shareholders. Other stakeholders such as authorities and social action groups can be considered in general, but not in particular, because the allocation is not yet known. The main question in this phase is: Does this innovation add more value to the stakeholders? Or reformulated: What objectives have to be realized by this innovation so that more value is added for all or nearly all stakeholders. The method to determine the different interests of stakeholder is described in Section 3.6.

Generally, process intensification leads to strong reductions in energy consumption and capital investments. From that perspective, it will serve the interests of customers and shareholders. However, other benefits also have to be considered. For example, an increase in product purity will strengthen the position in the present market and can also open new markets. Another example is that a possible reduction in waste will have both a financial and environmental benefit. The message is: Understand the interests of all (relevant) stakeholders and try to realize as much benefits as possible for them.

The importance of meeting SDGs cannot be underestimated. It is not only about technology and specifications, but it is also about sustainability and justice. On top of that, it is about image and the meaning of the brand. It can be expected that in the (near) future, shareholders will turn away from industries that excel in polluting the environment, producing $CO_2$, and increasing opposites in the world. If that's the case, it will be much more difficult to raise capital. Additionally, "green" companies will also move away from suppliers that are not "green enough." Paying attention to SDGs is not only about a sustainable and prosperous world but also about stakeholders and shareholders that support or not support a business. So, it is about the future of the company.

### 7.5.3 Beliefs in Society

Beliefs in society refers to the cultural and religious beliefs of the society in which the plant with the process-intensified process will be allocated. In this stage, the allocation is not yet decided. So, it is not possible to investigate the beliefs in society. However, there are two "buts."

The first "but" is that from the beginning, the allocation is clear, e.g. in case of a revamp. In that case, the beliefs in society have to be investigated. The second "but" is that the region or country is known but not the exact allocation. In that case too, beliefs in society have to be investigated.

In Section 3.7., some clues are given to map the specific cultural-religious beliefs that may influence the course of the project. In this stage, only those beliefs are important that have an influence on the design of the intensified process.

## 7.6 Multidisciplinary assessment method

In Chapter 5 of this book, we have presented the multidisciplinary assessment method. The key message of the method is to integrate the different bodies of knowledge in every stage of the innovation process, especially the knowledge of the practices of R&D, OPS, manufacturing and marketing & sales (M&S) have to be assessed. It is important to do this assessment in the stage-gate of the concept phase too. It is essential that the results of the foregoing steps (from scoping to validation) are first discussed with the stage-gate disciplines of marketing and OPS in separate meetings. The main reason is to make the contribution from every discipline, explicit and sharp. After that, the different contributions can be discussed in a joint meeting.

A multidisciplinary assessment for this phase can be done in the following manner:

1) The project team discusses the results of the concept stage with a broad group of engineers of the R&D department. This meeting focuses on the technical feasibility of the concept. Apart from the possible benefits, the development risks, and the scaling-up are discussed.
2) The project team discusses the results of the concept stage with a broad group of managers and process engineers of the OPS department. This meeting focuses on the used feedstock, control of the process, safety, and so on.
3) The project team discusses the results of the concept stage with a broad group of colleagues of the M&S department. In this meeting, the market opportunities

for the intensified process are discussed in more detail. Specifically, the expected added value for the various customers is discussed.

4) The project team analyses the discussions with R&D, OPS, and M&S. Similarities in views, differences in opinion, and new suggestions are identified. A concept multidisciplinary assessment report is written.

5) The project team invites two or three colleagues of the R&D, OPS, and M&S to discuss the concept multidisciplinary assessment report. Differences in opinion and new suggestions have to be discussed. It should be noted that at the end of this meeting, differences of opinion will not disappear and that everybody agrees about the new suggestions. The objective, however, is to make the different arguments explicit and to discuss whether there are differences of opinion about the validity of an argument, or about the weight of the different arguments.

6) Based on this discussion, the project team finalizes the multidisciplinary assessment report.

## 7.7 Exercises

### Exercise 7.7.1
Define the process boundaries for a process involving the conversion of algae harvested from the sea to crude oil.
Define output streams and input streams.
Determine sizes of output streams and input streams.
Define function blocks to transfer input streams to output streams.
Apply the BTG-BTL process for the conversion step.
Put in other function blocks to complete the functional design.

### Exercise 7.7.2
Take a case from Chapters 13–17 and execute the process synthesis steps 1–6 of Section 7.2.3 for that process case.

## References

[1]    Seider, WD, et. al., Product and process design principles – synthesis, analysis, and evaluation, 3rd ed. 2010, John Wiley, Hoboken, 2010.
[2]    Harmsen, J, et. al., Product and process design driving innovation, De Gruyter, Berlin, 2018.
[3]    Harmsen, J, Industrial process scale-up – a practical innovation guide from idea to commercial implementation, 2nd revised ed., Elsevier, Amsterdam, 2019.
[4]    Douglas, JM, Conceptual design of chemical processes, McGraw-Hill, New York, 1988.

[5]   Demirel SE, Li J, and Hasan MF, Systematic process intensification using building blocks, Computers & Chemical Engineering, 2017, Oct 4(105), 2–38.
[6]   Tian Y, Demirel SE, Hasan MF, and Pistikopoulos EN. An overview of process systems engineering approaches for process intensification: State of the art. Chemical Engineering and Processing-Process Intensification, 2018, Nov 1(133), 160–210.
[7]   Verloop, J, Insight in innovation, managing innovation by understanding the laws of innovation, Elsevier, Amsterdam, 2004.
[8]   Stankiewicz, A, van Gerven, T, and Stefanidis, G, The fundamentals of process intensification, John Wiley & Sons, Hoboken, 2019.

# 8 Feasibility stage

## 8.1 Introduction

The purpose of the feasibility stage is to determine the feasibility of the commercial-scale process for all aspects critical to the final successful implementation so that at the stage-gate, a decision can be taken to go to the costly development stage or not.

To that end, the feasibility stage also includes the making of a development plan to make an estimate of the associated cost.

Figure 8.1 shows the steps to be taken in the feasibility stage and the methods assisting these steps. The steps are described in Section 8.2 and the methods are provided in Sections 8.3–8.6.

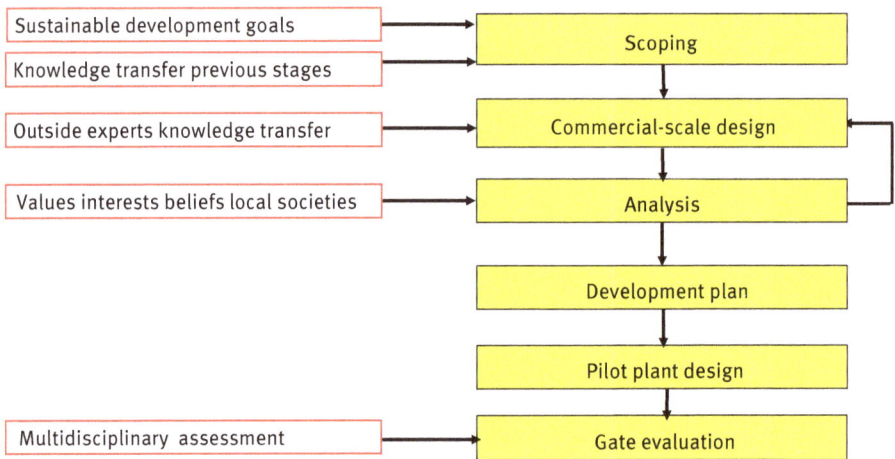

**Figure 8.1:** Feasibility stage: steps and methods.

### 8.1.1 Scoping

In this scoping step, all requirements for the end results of the feasibility stage are set. This involves requirements for the commercial-scale design, requirements for the development plan, and requirements for the pilot plant. The requirements for the commercial-scale process comprise:

- First, setting requirements for safety, health, environment, economics, technical feasibility, social, and contributions to sustainable development goals (SDGs) [1]. A special method for selecting the appropriate SDGs is provided in Section 8.2. Other requirements such as legal are also to be defined.

https://doi.org/10.1515/9783110657357-008

- Second, the scope states that the required process capacity or the process capacity and its related investment cost, have to be explored in combination with a market size study.
- Third, the scope states the country location or which country location options have to be analyzed in view of:
  - SDGs
  - Values of engineers perspective including knowledge, experience, and labor ethos of local process operators
  - Interest of stakeholders, that is, availability and transport cost of feeds
  - Local beliefs and political stability of the country perspective
- Fourth, the scope sets all product requirements from the clients' perspectives.
- Fifth, the scope sets all input specifications, the suppliers of these inputs, and requirements for the whole value chain, in view of circular economy and SDG. Section 7.2 is devoted to achieving the proper setting of the selected SDGs.

The requirements for the development plan are twofold. A list of knowledge that is lacking is to be obtained and a plan to generate that knowledge has to be made. This lack of knowledge identification requires a special process designer for the commercial-scale process. Each time he cannot make some design part because of lack of knowledge, he writes down that lack of knowledge and states an assumption to be able to continue with his commercial design or he changes the design such that this specific knowledge is not needed. He may have to make these types of decisions several times and, therefore, has to be experienced in this type of design. Hence, scoping also includes a statement on the requirements for the process engineer involved.

Scoping the downscaled pilot plant design includes its purpose. This can be validating the process design and its models. It can also involve making product samples for product testing.

Scoping also sets requirements for the product to be manufactured. It may include the country of choice for the commercial-scale process. Or, it includes choice of the part of the world for the process location and the wish to determine the best country inside that world section for the process location. So, either this choice has already been made by top management or they want to explore several options and if so, it is part of the design effort in this stage.

This scoping step is of enormous importance for the end quality of the feasibility stage. To that end, we advocate a multidisciplinary team from all relevant departments of the company to assist this scoping step. A multidisciplinary team method has been developed to guide the selection of team members and the scope setting. It is found in Section 7.3. The following story illustrates the importance of this multidisciplinary team approach in the scoping step.

### 8.1.1.1 Story lack of communication between disciplines on commercial-scale design

In the company I worked for a commercial-scale process was envisaged for a novel polymer. The process capacity was set and the commercial-scale design was made. The desired design and its related capital cost estimate were set at a high accuracy (10%) because top management wanted to take a decision based on a capital investment with low uncertainty. The design development took several months and considerable money was spent. The design and the economic figures, including the return on investment and the total investment, were presented to top management. They decided that they did not want to invest this large sum of money for this first-of-its-kind commercial-scale process and asked for a new design with a factor 4 lower investment cost. The new design was made for a much smaller capacity, again taking several months and presented to the top management. The investment cost was now indeed a factor 4 lower but the return on investment had also dropped considerably. The management then decided that they wanted the return on investment at the same level as the first design while keeping the total investment cost at the factor 4 lower. This meant that the whole process design had to be reconsidered and a new design base had to be found. After more than a year, this new base was found by modifying an existing process of a different product for the new polymer. So, million euros and more than 2 years were wasted because of lack of communication between top management and the engineers upfront of the commercial-scale design.

### 8.1.2 Commercial-scale design

The commercial-scale design is an essential part of the feasibility stage. It involves generating a complete process design with all equipment, connections, and auxiliaries. Also, all output streams will have to be defined in composition, size, and destinations (clients). Also, all input streams have to be defined in composition, size, and resources (suppliers).

Commercial-scale design starts with studying the scope description. This can be a thorough analysis of what the scope means for the design and then starting to design. It can also be a quick glance of what is needed, intuitively select the most critical requirement, i.e. the hardest to meet requirement, make a provisional design followed by checking the design result with the scope requirements, and modify the design until all requirements are met. It is up to the designer on how he approaches the design. The experienced designer will have his own way of approaching the design generation. For young engineers, the second approach is easier. It means, choosing the hardest-to-fulfill requirement first, design to fulfill the requirement, and then further design to meet the other criteria. In any case, the designer will cycle between synthesis and analysis.

What can also happen is that the designer has to make an important decision when nothing is stated in the scope that is relevant to that decision. In general, design scopes are under-defined but in the feasibility stage design, this is more likely to happen than in previous stages, as the design has to be comprehensive. If that is the case, the designer can convey to the management team his design decision dilemma and ask for an additional requirement in the scope to resolve the dilemma.

Here are some general design heuristics for meeting economic, safety, health, and environmental criteria.

### Heuristic A: Choose the reaction system with the highest selectivity to the product

Following this heuristic means that the lowest feedstock cost per ton of product is obtained and that the lowest number of separations is also likely to be obtained; hence, lower investment cost and lower energy requirements. This, in turn, means lower global warming gas emissions. The industrial case of Chapter 15 shows the advantages of applying this rule.

### Heuristic B: Do not consume additional chemicals for reactions or separations

Following this heuristic means, first, the lowest feedstock cost and, second, no additional waste streams are to be treated.

### Heuristic C: Do not use stripping with nitrogen or washing with water

Following this heuristic means that no additional waste streams are created which need further end-of-pipe treatment. Alternative stripping gases and washing liquids can be found by considering each process stream, including feed and product streams.

The designer will, furthermore, take the process concept design of the previous stage as the starting point and determines what needs to be added in auxiliaries and storage facilities. He may also reconsider the decisions made in the concept design in view of the more elaborate scope or because he is a more experienced designer can estimate the technical feasibility of the concept design better.

The designer will, furthermore, study whether process will be placed in an existing process site, a so-called brown field, or in a green field with no facilities. For a brown field, he will find out whether the higher demands for utilities can be accommodated without additional investment. If not, he will obtain an estimate of the additional cost involved. If the new design is in a green field, the designer may decide to list the utility requirements without designing them.

The amount of detail required for the commercial-scale design is, in general, such that at least a 30% accurate design can be made. This means that all equipment inside the battery limits are specified in type and size. Equipment located outside battery limits such as all utility cost (steam, electricity, nitrogen, waste water treatment) are quantified in amount and a cost estimate is made based on historic cost, keeping in mind if it is a brown field or a green field project.

All major PI design decisions, probably, have been made in the concept stage. However, these decisions have to be reconsidered in the feasibility stage design in view of the scope and their feasibility for commercial-scale design. This means that the PI process technology options selected will also be revisited. The PI methods of the concept design can be applied but the scalability aspects to be considered in the PI method are provided in Section 8.4.

This commercial-scale design is not a simple linear process but cycles several times between design synthesis and analysis. This also means that the supporting methods are applied several times. The next section describes the design analysis elements.

### 8.1.3 Analysis

#### 8.1.3.1 Analysis to meet scope requirements

Design synthesis and analysis are not strictly sequential steps but have many re-cycles, from analysis back to synthesis. The commercial-scale design will be analyzed if it meets the design scope. If that is not the case, it will be modified in the synthesis step. The design will also be analyzed whether it is optimal or sub-optimal. It will then be optimized by process modeling, simulation, and design changes. Thus, this is also a cycle between analysis by simulation and synthesis. For process intensified processes, this modeling is a far from trivial effort as standard computer simulation modules for PI designs are, in most cases, not available and considerable modeling effort is required. This is particular so for designs involving PI domains: function synergy, temporal, and thermodynamic (alternative energy supply).

The analysis may also concern Values of Engineers perspectives. At first instance, this may be hard to understand because these values are not easily seen. However, these values play an important role. This is revealed in the Eastman case described in Chapter 14.

This case shows the importance of the commercial-scale design in the feasibility stage by which, the seriousness of the commercial-scale design puts the engineers in the spotlight, triggering the assessment based on their values of having a reliably easy-to-operate process. A method for an engineering value analysis is described in Section 7.5.

The analysis can also be varied out from an Interest of stakeholders perspective. An example is finding a relationship between the local stakeholders and the process design and/or SDGs so that the design is embedded in the local society. In cases where various countries are considered for locating the process, this interest of stakeholders perspective becomes even more important. Section 7.6 provides a method for this stakeholders analysis.

The analysis may also be on the relevance of the local society beliefs perspective to the design. The country of selection can depend on the local dominant beliefs of the countries envisaged. Also, this element can be considered for a decision on which country to choose for the process location. A method for this beliefs analysis can be found in Section 7.7.

**Story of ethylene oxide process design and local stakeholders involvement Shell Canada**
Chapter 15 shows the conventional ethylene glycol process of which the ethylene oxide process is a subsection. The reader may want to look at the process scheme for a better understanding of the text below.

The director of Shell Canada wanted to incorporate sustainable development into Shell Chemicals Canada. To that end, he set up a Sustainability Advisory Panel. Twelve representatives of the local community where asked to be part of that panel to comment regarding social, environmental and economic impacts on six processes. If a representative did not like an aspect of the design, he would put up a red flag and state his objection [2].

This panel evaluated a very large-scale ethylene glycol process to be implemented in Scotford, Canada. One of the panel's objections of the process design was the very large emission of carbon dioxide, a by-product of the ethylene oxide reaction step. This carbon dioxide is relatively pure. Another member of the panel then proposed to use this carbon dioxide for the vegetable production in green houses near the process location. Shell then investigated the requirements for using the carbon dioxide stream for this purpose. It appeared that the carbon dioxide should be completely free of ethylene as this acts as a ripening plant hormone. An additional process step to remove trace amounts of ethylene was added to the process design. A pipeline was also planned to be laid to the greenhouse farmers. It appeared that the modified design also had a higher return on investment. The modified design was accepted by the stakeholder panel and the process was constructed and implemented [2]. Hence, stakeholder involvement can lead to surprisingly positive results.

### 8.1.3.2 Analysis for lack of knowledge
Several times during the commercial-scale design, the designer will stumble to make a design decision due to lack of information and lack of knowledge. He may change the design so that he does not need this piece of knowledge. Or, he may state the specified lack of knowledge and that it should be addressed in the development plan. By making an assumption, he will then continue to design.

### 8.1.4 Development plan

A good starting point for making a development plan is listing all items where sufficient knowledge is not available for a reliable commercial-scale design as mentioned by the process designer, see previous section. Then, for each item, a specific plan is defined to obtain this knowledge. The collective of these plans are then part of the development plan. This is often called the scale-up plan. For a large number of PI technologies, scale-up methods are provided in Chapter 2. Further, guidelines-integrated process scale-up are provided in the industrial process scale-up book by Harmsen [3].

However, additional plans should be made for risk reductions associated with values of engineers, interests of stakeholders, and risks associated with local beliefs. Sections 7.5–7.7 will be helpful for making these plans.

### 8.1.5 Pilot plan design

If an integrated pilot plant is needed, it will be designed as a downscaled version of the commercial-scale plant. Also, a test program will be made with an estimate of total time and effort required. Finally, the pilot plant investment and the testing cost will be estimated. Harmsen provides details on when an integrated pilot plant is needed, how it is designed, and what needs to be tested. [3]. Hence, in the feasibility stage, a pilot plant design is made.

### 8.1.6 Stage-gate evaluation

The results of the feasibility study will have to be reported in writing and presented to the stage-gate panel.

The report will contain, at least for the commercial scale, the feasibility stage results keeping in mind the scoping requirements. These will include the Safety, Health and Environmental aspects concerning the commercial-scale design:

The economics in terms of investment, return on investment, variable cost, and fixed cost

The technical feasibility showing that Technology Readiness Level 5 is obtained [1].

The quantified contributions to selected SDG

Evaluations of selected Process Intensified technologies

Evaluations of Values of Engineers (including capability of developing the process to commercialization)

Evaluations of Interests of all relevant Stakeholders

Evaluations of Beliefs of the local society where the commercial-scale process is to be placed

The last three evaluations are based on the VIB method described in Chapter 4. The report will, furthermore, contain the development plan:

A development plan listing all the identified lack of knowledge that are needed for a pilot plant design and test program.

The total cost and a justification for the costs.

It may contain a risk table containing the risks, with and without the development plan.

The stage-gate evaluation will be executed by a multidisciplinary team panel. A method for guiding this evaluation is described in Section 8.6.

## 8.2 Sustainable development goals method

The end product of the concept stage is a process design containing all essential technologies with all critical aspects for commercial-scale implementation. This process design involves an experimental validation and a multidisciplinary assessment.

In the scoping of the feasibility stage, the process design will be reviewed again from the perspective of the SDGs. Chapter 3 presents a description of the different SDGs. In this stage, the review can be more concrete because "all essential technologies" are known and "all critical aspects for commercial-scale implementation' are identified. The main objective of this stage is to mirror the requirements of the SDGs to the process design.

In this stage, the discussion can be in two rounds. In the first round, every SDG is discussed briefly. The description provided in Section 3.2. is taken as a starting point. The main question is, does this SDG offer essential requirements for the concept phase? For example, SDG 12 (Ensure sustainable consumption and production patterns) and SDG 13 (Take urgent action to combat climate change and its impacts) can be very important to define the targets in the feasibility stage. They challenge engineers to think about the whole chain and set targets for the circularity of the materials flow, the amount of waste materials, the purity of the material, the energy consumption, the $CO_2$ production, and so on. The first round will result in a number of SDGs that have the potential to contribute to the targets set for the intensified process.

In the second round, every selected SDG is discussed in more detail. The sub goals of the SDG, as presented in *The 2030 agenda,* will support the engineers to identify the most important parameters and set the targets for the intensified process. In this round, three types of outcomes are possible for every individual SDG:

1) The present process design does not meet the standards.
2) The present process design does meet the standards.
3) The present process design does surpass the standards.

At first, we would like to make a general remark about the SDGs. They define the targets for the year 2030. However, every PI project that will be started now – considering the project time and the long writing-off times of new equipment – will cover a long period beyond 2030. In other words, the project team is invited to discuss not only the 2030 targets but also to consider whether, in the future (beyond 2030), more severe targets can be expected. As a general rule, it can be stated that the targets for energy use, emissions, waste materials, and circularity will increase.

We return to the evaluation of individual SDGs. All SDGs with the outcome "does not meet the standards" have to be problematized. For example, by identifying a lack of knowledge and may be by making this SDG explicit in the development plan or the pilot plant design. All SDGs with the outcome "does meet the standards" and "does surpass the standards" are unproblematic. That means that these standards will be met by the proposed process and no problems can be expected in the future.

## 8.3 Knowledge transfer previous stages

### 8.3.1 Introduction

The Spanish philosopher and essayist George Santayana (1863–1952) coined the famous statement: "Those who cannot remember the past are condemned to repeat it." This statement has produced many paraphrases and variants. One of the most striking is: "Those who do not remember their past are condemned to repeat their mistakes." This variant is surely relevant for innovation projects in the process industry.

In any innovation project, the road from the discovery stage up to and including the start-up stage is a bumpy one with many uncertainties and surprises. In formal documents like reports and specifications, the most important information and knowledge generated is made explicit.

However, engineers also have generated a lot of "tacit knowledge" about a newly developed process. That kind of knowledge is difficult to transfer to another person by writing or by verbalizing it. It is a kind of knowledge that is in the "heads" and the "fingers" of the engineers. Often, they are not even aware that they have that knowledge and that it is an important knowledge. One of the most important challenges of innovation projects is "to remember the past" in order to prevent that "their mistakes are repeated." If, however, this tacit knowledge is not transferred, then a project can fail.

An example of such a failure is provided here. A new fermentation recipe was developed in Shell Bioscience Laboratory, Sittingbourne, UK in the eighties of the last century. I worked there at that time as a research technologist. A fermentation recipe for a new product was developed and then transferred for commercial-scale production in an existing fermenter of a different fermentation company. Despite many attempts, fermentation did not happen. In the end, the researcher who developed the recipe, Jan Drozd, was invited to visit the manufacturing site and watch the whole procedure. The first step was sterilizing the batch content. Jan then heard a loud and strange noise. He asked: "What is that noise?" The operator explained that it was live steam being blown into the fermenter to sterilize the broth. Jan was astonished. In his laboratory, he had chemically sterilized the broth. He figured that by blowing live steam through the fermenter, ammonia, an essential ingredient for microorganism growth, would be stripped from the liquid. He then proposed that his chemical sterilization method be followed. He accepted that it was his fault that the method was not part of the originally provided recipe description. The problem was solved by the now explicit sterilization method [3].

There are four complementary ways to prevent repetition of mistakes and failures of the past during the development of a PI innovation:
1) Having the same people in all innovation project stages
2) A well-considered team composition during the various stages, for continuity in knowledge and experience.
3) Dialogue sessions inside project team, in which the learned lessons of the past are shared with new team members.
4) Dialogue sessions with experienced engineers and managers from the mother company, and from equipment and knowledge suppliers.

The first way is impossible to practice as each innovation stage requires its own specialist knowledge and also that process innovation projects can easily take 5 years or more. This method, however, can be partly applied by having one or two employees traveling with the project through all stages, as shown in the BTG-BTL case of Chapter 13.

Ways 2–4 are explained in the next sections.

## 8.3.2 Well-considered team composition

This way focuses on time travel of knowledge by knowledge carriers in the team. In modern technology, it is believed that engineers are rational beings. They use rational knowledge, they have rational dialogues, and they make rational decisions. This view is quite attractive because it focuses on the strength of engineers: their rationality. It also emphasizes the key capacity of engineers: to gather all relevant data for designing new processes. However, for more experienced engineers, this belief is not obvious. They have experienced many times in their career that "not all knowledge is rational" and that it is not evident that "all relevant data are gathered." In other words, rational data and rational procedures are embedded in the experience of engineers. The richer the experience of the engineer, the richer the meaning of rational knowledge. In other words, all knowledge is human knowledge.

This statement is more true in the development of intensified processes. All cases show that knowledge is human knowledge. The art of process intensification also requires the art of connecting people. This means, connecting people with different disciplinary backgrounds, different experiences, and different organizational roles.

A key condition in every phase (stage) is the composition of the project team. We would like to make several general remarks. First, in every phase, there have to be members with a background in R&D, Operation Services (OPS), and Marketing & Sales (M&S). In other words, the barriers between the various disciplines must be removed from the start. Second, in all phases, there has to be a continuity in knowledge carriers with respect to R&D, OPS, and M&S. It forms the backbone of the process innovation project. On the one hand, they form the memory of the project and on the other hand, they give stability to the project and the project team. Third, in every phase, new members will come in. These members have at least two functions: bring in new knowledge that is required for the relevant stage and critically discuss the present findings and proposal using their disciplinary knowledge and their experience. Finally, social investments have to made (meaning, meeting time together) to make sure that the new team will develop as a team.

Another key condition in every phase is the team for the multidisciplinary assessment. It is proposed that that this assessment is carried out not only by the different disciplines within the team, but that experienced experts from R&D, OPS, and M&S are invited to discuss the findings of the project team. Every organization and every discipline has members who are known for their knowledge and (super) critical view. Generally, these members are stubborn, self-confident, and highly motivated, although their manners could often be more civilized. However, they are of utmost importance for the organization.

It should be noted that the cost of failure in process innovation in the end is so high that any cut in the team's knowledge transfer is out of question. Proofs of the statement are found in innovation failures due to lack of knowledge transfer documented in an Independent Project Analysis by analyzing 12,000 process innovation projects and summarized by Harmsen [3].

### 8.3.3 Time Travel Dialogue sessions with new team members

This way focuses on knowledge that has been acquired in previous stages of the project and that needs to be transferred to the feasibility and follow-up stages. This method is called Time Travel Dialogue sessions. They have to be organized with the new team in which the foregoing stages of the project are "repeated" in a kind of pressure cooker process.

The following analyses of the foregoing stages have to be repeated in these sessions
- The analysis of the SDGs;
- The PI design method
- The VIB analysis
- The multidisciplinary assessment

### The analysis of the SDGs
The following procedure can be used to repeat the analysis of the SDGs in a pressure cooker process:
1) The analysis is done per every individual SDG (see Chapter 3).
2) The new team discusses the importance of SDGs relevant for the project. Especially, new members are invited to present their view.
3) After that, the discussion leader presents the final judgment of the foregoing stage(s) with respect to the relevant SDGs.
4) The differences between the views of the new team (step 2) and the results of the foregoing stage(s) (step 3) are discussed. The discussion may lead to changes in the view on the importance of the relevant SDG for the intensified process.
5) Steps 2–4 are repeated till every SDG has been analyzed.
6) The analysis is closed by a joint discussion of the whole team about SDGs and the importance of these results for the present phase (stage).

### The PI design method
The following procedure can be used to repeat the PI design in a pressure cooker process:
1) One of the engineers of the R&D department presents a short introduction of the present process and the PI principles of Chapter 2.
2) The discussion leader invites the new members of the team to reflect on the question – how the present process can be improved by the application of the different concepts of PI.
3) Then, one of the engineers of the R&D department tells the story, starting from the discovery stage up to the foregoing stage, about the design of the new process, which considerations played a role, which experiments have been done, which dilemmas came up, and what decisions were made.
4) The new members are invited to reflect on the presentation about the intensified process.
5) The reflection is closed by a joint discussion of the whole team about the PI design method and the importance of the results for the present phase.

**The VIB analysis**

The following procedure can be used to repeat the VIB analysis in a pressure cooker process:

1) One of the engineers presents a short introduction of the VIB method (see Chapter 4).
2) The discussion leader invites the new team to reflect on the values of engineers. Especially, the new team members are invited to air their views.
3) The discussion leader presents the conclusion of the foregoing stage(s) stage with respect to the values of engineers.
4) The discussion is closed with an evaluation of the differences between the views of the new team (step 2) and the conclusion of foregoing stage(s) (step 3) by the whole team.
5) The discussion leader invites the new team to reflect on the interests of stakeholders. Especially, the new team members are invited to air their views.
6) The discussion leader presents the conclusion of the foregoing stage(s) with respect to the interests of stakeholders.
7) The discussion is closed with an evaluation of the differences between the views of the new team (step 5) and the conclusion of the foregoing stage(s) (step 6) by the whole team.
8) The discussion leader invites the new team to reflect on the beliefs in society for the relevant allocation. Especially, the new team members are invited to air their views.
9) The discussion leader presents the conclusion of the foregoing stage(s) with respect to the beliefs in society for the relevant allocation.
10) The discussion is closed with an evaluation of the differences between the views of the new team (step 8) and the conclusion of the foregoing stage(s) (step 9) by the whole team.
11) The analysis is closed by a joint discussion of the whole team about the results of the VIB analysis and the importance of these results for the present phase.

**The multidisciplinary assessment**

The following procedure can be used to repeat the multidisciplinary assessment in a pressure cooker process:

1) One of the engineers presents a summary of the multidisciplinary assessment of the foregoing phase(s).
2) The topics on which the members of the multidisciplinary assessment agreed are discussed. Especially, attention has to be given to the underlying points of their agreement.
3) The topics on which the members of the multidisciplinary assessment did disagree are discussed intensively. Especially, attention has to be given to the arguments of

different disciplines, how the different arguments were weighed, which decision was made in the gate evaluation, and the substantiation of the decision.

4) The analysis is closed by a joint discussion of the whole team about the foregoing multidisciplinary assessment(s) and the importance of these results for the present phase.

## 8.4 Outside experts' knowledge transfer method

This method focuses on knowledge and experience transfer of people outside the project, such as senior engineers and senior managers in the own company, and important suppliers of equipment and knowledge institutions in the field of PI. Their knowledge and experience can be explored to review the state of the art commercial-scale design at this feasibility stage.

This assessment of the preliminary commercial-scale design with the outside experts can be done as follows:

1) The project team discusses which disciplines have to be part of the multidisciplinary assessment. At least the departments R&D, manufacturing, OPS, and M&S have to be present and also other internal and external specialists have to be invited.

2) The project team presents the results of the commercial-scale design with external specialists in a specific discipline. This is done for each discipline. The meeting focuses mainly on the disciplinary knowledge of the attendants. The most important question is: How do you judge the state of the art of this stage? What are the opportunities? What are the risks? This can be done with the individual disciplines related to R&D, Manufacturing, and OPS. The project team analyses the results of the individual meetings with the different disciplines. Similarities and differences in opinion have to be identified. Especially, differences of opinion have to be highlighted. A concept assessment report is then written.

3) The project team invites two or three attendees from every discipline to discuss this concept multidisciplinary assessment report. First, the agreements in opinion have to be reviewed briefly: What is the importance of these agreements for the gate evaluation? Then, the differences of opinion have to be discussed: What is the impact of these disagreements for the gate evaluation? Preferably, the discussions about these difference in opinion should lead to a) agreement about the issue, b) the agreement to disagree about the issue, and c) the recognition that more research is required to judge the issue.

4) Based on this discussion, the project team finalizes the multidisciplinary assessment report. It has to be noted that differences of opinion that could have an influence on the gate evaluation have to be made explicit in the report.

An example of the power of a multidisciplinary assessment is found in the OMEGA case of Chapter 15, where the marketing and sales experts of Shell received a wish from their customer to have the Mitsubishi glycol process technology delivered. Sales then asked engineers of Shell to evaluate the Mitsubishi process. They concluded that it was indeed an excellent process technology but it had to be improved for specific items.

## 8.5 Design analysis VIB method

The application of the Values of engineers, Interests of stakeholders, and Beliefs of society (VIB) method as described in Chapter 3 supports the analysis step of this stage. The main reason for that is engineers are invited to discuss their process proposal and the model from the different perspectives of the VIB model. In fact, engineers are invited to go one step back in order to discuss the process proposal and model from the perspective of the engineers, the perspective of stakeholders, and the perspective of the beliefs in society.

### 8.5.1 Values of engineers

The process proposal and the model are "confronted" with the world of the engineer. They are urged to answer the question again: What values are important to me? Or: Which values drive this PI project? It should be noted that in comparison to the foregoing phase, a much larger group of people are involved in the project. Additionally, a lot of new information has been gathered. For that reason, the values of engineers have to be discussed explicitly in this phase. In hapter 4, a method is proposed to reflect upon values. In addition, a general list of engineering values is also provided. Together with the results of the SDG method, a more complete and more focused list of engineering values will evolve.

In this stage, it is also important to discuss the specific meaning of every value for this PI project. For example, what is the specific meaning of the word "safety" in this project? It implies to answer questions such as: "What are the safety issues in the present process?" "What are the safety issues of the intensified process?" "How to improve the performance of the intensified process with respect to safety?"

### 8.5.2 Interests of stakeholders

In this stage, "all essential technologies" are known and "all critical aspects for commercial-scale implementation" are identified. Also, the most important stakeholders are identified. The main question in this analysis is: Does the proposed intensified process add value to the stakeholders? Or reformulated: What objectives have to be realized by this PI project so that value is added to all stakeholders. Section 3.6 offers a general method to determine the different interests of stakeholders.

Individual stakeholders will be invited so that the purpose and content of the project can be explained. Special attention will be paid to the SDGs to which the project contributes. Stakeholders can state their wishes and these can then be incorporated into the development plan.

In the feasibility stage also a proposal will made for the allocation of the activity. Generally, it is a part of the scoping process to identify various possible allocations. For every possible location, an inventory of the relevant stakeholders has to be made. As far as possible, the method provided in Section 3.6. has to be used to determine the different interests of these stakeholders.

### 8.5.3 Beliefs in society

Beliefs in society refer to the cultural and religious beliefs of the society in which the plant with the process intensified process will be allocated. In scoping, the various options for allocation are identified. In this stage, these different options can be evaluated with respect to the item "beliefs in society."

In Section 3.7, clues are given to map the specific cultural-religious beliefs that may influence a PI project. The examples offer a guide to investigate the local beliefs in society. In case the combination of the PI process and the proposed country of allocation raises a lot of questions, it is recommended to visit the relevant country to investigate these beliefs in more detail.

## 8.6 Multidisciplinary stage-gate assessment method

In Chapter 5 of this book, we have discussed the multidisciplinary assessment method. It is a twofold method. First, to identify and highlight the specific knowledge of the different departments and/or the different hierarchical levels of the organization. Second, to integrate the different bodies of knowledge. In this phase, the knowledge of the practices of R&D, OPS, and M&S have to be assessed.

Third, the design will need to be assessed with a long list of requirements regarding, at the minimum, the safety, health, environmental, economics, technical feasibility, and sustainable development contributions (SHEETS list).

A multidisciplinary assessment for this phase can be carried out in the following way:
1)   The project team discusses the results of the feasibility stage with a broad group of engineers of the R&D department. This meeting focuses on the technical feasibility

of this project. Among other possible benefits, the development risks and scaling-up issues are discussed.

2) The project team discusses the results of the feasibility stage with a broad group of managers and process engineers of the OPS department. This meeting focuses on the used feedstock, control of the process, safety, and so on.

3) The project team discusses the results of the feasibility stage with a broad group of colleagues of the M&S department. In this meeting, the market opportunities for the intensified process are discussed in more detail. Especially, the expected value added to the various customers is discussed.

4) The project team discusses the results of the feasibility stage with the safety, health, and environmental (SHE) department. In this meeting, these aspects for the intensified process are discussed in more detail. Especially, it is important to identify in which aspects the intensified process is better than the commercial-scale reference process. It goes without saying that the "aspects in which the intensified process is worse will require additional actions in the development stage to mitigate the safety risks."

5) The project team analyses the discussions with R&D, OPS, M&S, and SHE departments. Similarities in views, differences in opinion, and new suggestions have to be identified. Especially, attention has to be paid to differences in opinion. A concept multidisciplinary assessment report is written.

6) The project team invites two or three colleagues of the R&D, OPS, M&S, and SHE departments to discuss the concept multidisciplinary assessment report. Especially, the importance and the impact of differences in opinion have to be discussed. Preferably, the discussions about these difference in opinion should lead to (a) agreement about the issue, (b) the agreement to disagree about the issue, and (c) the recognition that more research is required to judge the issue.

7) Based on this discussion, the project team finalizes the multidisciplinary assessment report. It has to be noted that differences in opinion that could have an influence on the gate evaluation have to be made explicit in the report.

## References

[1] Harmsen, J, Product and process design driving innovation, Degruyter, Berlin, 2018.

[2] Scholes, G, Integrating sustainable development into the shell chemicals business, EFCE Event 616, Delft University Press, Delft, 1999, 69–82.

[3] Harmsen, J, Industrial process scale-up – A practical innovation guide to idea to commercial implementation, 2nd revised edition, Elsevier, Amsterdam, 2019.

# 9 Development stage

## 9.1 Introduction

The purpose of the development stage is to obtain a proven design on which a decision about entering the commercialization stage can be taken in the stage-gate to engineering procurement and construction (EPC). In the development stage, the commercial-scale process design of the feasibility stage will be validated through pilot plant tests. In some cases cold flow model tests are also used to validate scale-up. Also, the commercial-scale design of the feasibility stage will be improved with the results of the tests. It will be further defined in more detail in the front-end engineering design. All steps of the development stage, as shown in Figure 9.1 will be described in the following sections. The supporting methods shown in Figure 9.1 will be described in Sections 9.8 and 9.9.

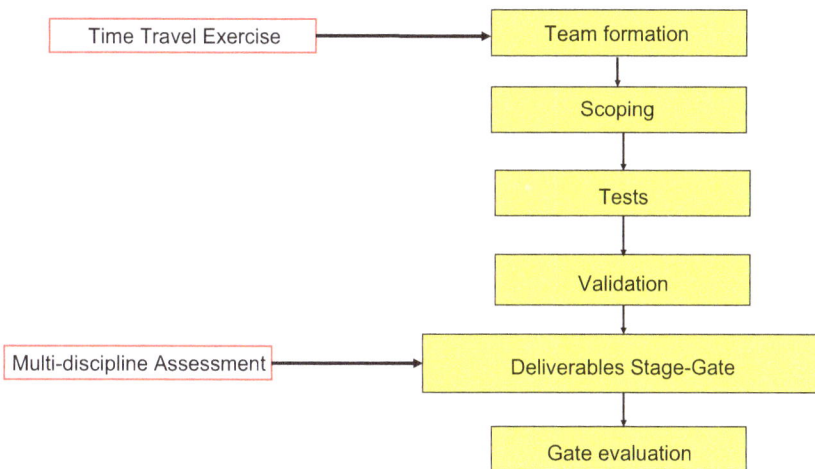

**Figure 9.1:** Development stage: steps and methods.

## 9.2 Team formation

An important element is the composition of the process development team. For large process development, it will require process operators and process engineers. One of the latter is often the team leader. Furthermore, the team will need supporting specialists such as analytical chemists and materials construction experts. To familiarize these employees with the knowledge generated in the previous stages

https://doi.org/10.1515/9783110657357-009

and make it their own project, a specific method called time travel exercise is provided in Section 9.8.

This team will also have to generate risk items not identified in the feasibility stage, as this team is focused on what can go wrong. The scope should contain a statement that all risks should be identified and treated in the development stage.

A method to build the team knowledge from previous stages, a supporting method, is provided in Section 9.8.

## 9.3 Scoping

In scoping the development stage, all the requirements to obtain the project results are set, based on which the decision for the investment for the commercial-scale plant can be taken with confidence. The development stage is very different from the previous stages. Now, the focus is on making sure that all risks are reduced to a very low level and that the process is defined in sufficient detail, and is assessed for the decision to the next stage: the engineering procurement, construction, and start-up.

The scope will have the same requirements as the feasibility stage as well as new requirements such as that of a trusted scale-up method and a complete set of deliverables for the next stage – engineering, procurement, and construction – and for the stage-gate decision. That set is provided by Harmsen [1].

## 9.4 Tests: defining development plan

In general, the development plan for a new process has the following elements:
– Validation of the process design and model, in most cases by a pilot plant
– Scale-up method selection and validation
– Front end engineering design
– Assessment of remaining risks and benefits for commercial-scale implementation

Defining this development plan is carried out by the development project leader, an experienced process developer. He will consult experts on certain aspects such a choice of construction materials and their testing, reaction engineers for scale-up methods; he will also consult the technology provider of the PI technology on details of the down-scaled pilot plant and scale-up knowledge and method. He will also organize the time travel exercise as described in Section 9.8 so that development team owns the project.

## 9.5 Validation

### 9.5.1 Validation process design

The process design made in the feasibility stage needs to be validated on its assumptions, the known unknowns and unknown unknowns. For the latter, a down-scaled pilot plant version of the commercial-scale process design containing all recycle streams will be needed. Harmsen provides details on when such a pilot plant is needed [2].

The process design should, in particular, be validated on the choices of construction materials, and on the long-term catalyst performance.

### 9.5.2 Validation process model

The process model on which the process design is based should also be validated. This validation should be carried out for the performance of each process unit and on the prediction of all stream compositions.

For function-integrated, process-intensified technologies, this validation should be carried out in detail for each function in the process unit; for instance, on the separation and reaction functions of a reactive distillation unit.

If the process-intensified technology is from the temporal domain and contains dynamic process- intensified technology, such as a reverse-flow reactor or a cyclic distillation column then the dynamic behavior should be compared with the dynamic model.

If the process-intensified technology uses alternative energy supply such as applying microwave energy, then not only the performance of the process unit but also the alternative energy supply model should be validated.

Furthermore, this validation should be carried out by long pilot-plant runs to see any unknown longer-term behavior, such as trace component build up causing foaming or corrosion- or catalyst decay.

### 9.5.3 Validation scale-up theory

Scale-up theory validation for the integrated process is covered in the previous sections on design and modeling validation. Scale-up theory validation for individual process units is dealt with, here.

Harmsen provides five scale-up methods available for process units [2]:
- Brute force scale-up
- Model-based scale-up
- Empirical scale-up

- Hybrid model-empirical scale up
- Dimensionless numbers (similarity) scale-up

A selection for each new process unit should be made and the selected scale-up theory should be validated.

### 9.5.3.1 Brute force validation

In the brute force scale-up, all critical success phenomena of the down-scaled version are retained by keeping the feed source and the hydrodynamic velocities the same. The only changes in scale-up are the diameter and the capacity. Special precautions are to be taken for the pilot plant to ensure that there are no wall effects and no heat exchange via the wall.

A variant of the brute force scale-up is the numbering up scale-up. In numbering up, the number of units increases with the capacity increase, but the individual process unit and its flow rates are kept the same.

This scale-up only holds if the feed flow distributions are kept uniform over the entire cross-sectional feed area. If this cannot be ensured, then other scale-up methods should be considered, as described in the next sections. A cold flow experimental setup can be used to validate the scaled-up distributor.

The advantages of this scale-up method are that it does not need detailed kinetic, mass transfer, and residence time distribution modeling. Also, communicating this scale-up method to management is relatively easy. The higher cost for the larger pilot plant is often accepted because of the low-risk scale-up method.

This scale-up method is particularly suitable for conversions involving complex feed materials, such as in crude oil refining and biomass conversions, as no detailed chemistry and kinetic knowledge is needed for scale-up.

An example of the brute force scale-up application is the reactive distillation of CDTECH in which a pilot plant of the same height, the same catalytic packing, and the same feed material is used as in the commercial scale. Special precautions are taken to insulate the wall so that the pilot plant is adiabatic. In scale-up, the flow rate and the diameter are increased such that the superficial velocities of gas and liquid are kept the same [2].

### 9.5.3.2 Model based scale-up validation

In model-based scale-up, all physical, chemical, thermodynamic and hydrodynamic effects on the process unit performance and also the parameters that change during scale-up are described. The model can then be validated by changing certain parameters in the pilot plant and comparing the experimental outcome with the model predictions. It is important that no parameters are introduced afterward to fit the model to the experimental outcomes. If the model deviates from the experimental outcomes, then the cause for such deviations should be analyzed and a new model should be made. If the new model also is not in agreement with the original model, then a different scale-up method should be selected.

The advantage of this model-based scale-up is that the pilot plant can be small, because the modeling optimization of the commercial-scale process (prior to designing the pilot plant) can be done for many design parameters, and a "robust" optimum can be found: "robust" in the sense that at that optimum, the performance is not very sensitive to variations in the parameters. Hence, a reliable design can be obtained and also a well-designed scaled-down pilot plant can be defined to validate the model.

Communicating the scale-up method to management is however more difficult than in the brute force scale-up. Reporting the model validation by pilot plant tests clearly will help convince management.

An example of model-based scale-up is reactive distillation by Sulzer for chemicals production. A small-scale pilot plant is used. Models for mass transfer, pressure drop, and kinetics are validated in this pilot plant. Upon scale-up, the models are used to design the commercial-scale plant in which the packing dimensions and column dimensions and velocities are increased [3].

### 9.5.3.3 Empirical scale-up validation
In empirical scale-up, the unit operation is experimentally tested at several scales. Trends on performance with scale-up are determined, leading to a prediction on the effect of scale. The prediction is then validated by an even larger unit operation. Due to the uncertainties surrounding this scale-up method, often three or four test scales are applied.

If the trend prediction fails after these three or four scales, then further scale-up with this method is concluded as being unreliable. Further scale-up is then achieved by having several units in parallel (numbering up) to obtain the commercial-scale production. I have noticed in one case, that 36 reactors in parallel were chosen to obtain the commercial-scale production.

This empirical scale-up is often applied for mechanically stirred reactors applied in polymerization, fine chemicals, and pharmaceuticals production. Often, the reactors are operated batch wise. Scale-up of mechanically stirred tank reactors where mass transfer and mixing phenomena are critical success factors is notoriously difficult. The batch-wise operation adds to the scale-up difficulty, as the time to reach a certain conversion will depend on these phenomena, and different reaction times, in general, means different by-product formation.

Process-intensified reactors for which a brute force scale-up method can be applied compare favorably with these mechanically stirred reactors. For instance, these can be static mixer reactors in which the static mixer elements are also the heat exchanger or milli-pore reactors. Chapter 2 provides an overview of PI technologies and their available scale-up methods.

An example of empirical scale-up and a microreactor scale-up is provided by Schwalbe. The empirical scale-up involves using a series of existing mechanically stirred reactors, including the commercial scale reactor. The microreactor scale-up method is not described by Schwalbe, but probably involves a brute force scale-up, by just increasing the number of microchannels. The mechanical reactor scale-up cost 10 man-days and resulted in a lower-yield product than that of the microreactor. The investment of the microreactor of 0.4M€ had a pay-back time of 0.2 year, mainly due to the higher yield of product on feedstock [4].

### 9.5.3.4 Hybrid model-empirical scale-up validation

In the hybrid model-empirical scale-up, experiments at a small scale are carried out and models that agree with the experimental results are made. A next scale experimental setup is tested, and the results are compared with the model. Further work may be carried out on to obtain agreement between model and results. Then, by a large-scale setup, the results are again compared with the model. If the model is then in agreement with the results, some confidence that further scale-up using the model is reliable is obtained.

A theoretical example to illustrate the method is a baffled oscillating flow column for crystallization, described in Chapter 2. The particle distribution results of a small-scale unit correlate with the energy dissipation and with the fluctuating Reynolds number, when changes to the oscillation frequency and amplitude are made. A larger scale setup is designed using the baffle geometry scaled with the column diameter. The particle size distribution is compared with the Reynolds and power dissipation correlations obtained from the small scale. The correlation that predicts the large-scale results is then selected for further scale-up.

### 9.5.3.5 Dimensionless number scale-up validation

In dimensionless number scale-up, all dimensionless numbers governing the process unit are kept at the same value. For simple geometries like tubes, the method is the same as the brute force scale-up method. For complex geometries and complex phenomena, the method is of little use, as one cannot be certain that all phenomena and geometry aspects have been considered; communicating the method is also impossible. The method is therefore not recommended.

### 9.5.3.6 Other scale-up methods

Vogel discusses scale-up using mini-plants in between the bench-scale and the pilot plant scale, or as a replacement of the pilot plant scale [5]. For classic unit operations not involving solids processing, this is a useful addition to scale-up methods; however, it is less applicable in process-intensified processes.

## 9.6  Deliverables for the stage-gate

A comprehensive list of deliverables for the stage-gate is obtained by combining a list by Bakker [6], and Harmsen [2]. Furthermore, items about Sustainable Development, Values, Interests, and Beliefs, and multidisciplinary assessment of this book are added. All the items of the list concern the commercial-scale process implementation. Here is the list:

1) Evaluation of all development results report
2) Front end engineering and design (FEED) report
3) Investment cost estimate report
4) Hazard and operability (HAZOP) report
5) Safety and health of operators and local society report
6) Environmental impact report
7) Economic report (or for some companies a business case report).
8) Sustainable development report
9) Values, interests, and beliefs report
10) Risk assessment development results and market survey results report
11) Planning of engineering, procurement, and construction effort and schedule
12) Process start-up plan
13) Multidisciplinary assessment report

## 9.7  Stage-gate evaluation

Stage gate evaluation will be on all deliverables presented in Section 9.6.

The stage-gate evaluation will be carried out by a multidisciplinary team of managers of at least marketing & sales, operation, safety, health and environmental (SHE) department, and R&D department. Often, maintenance engineering and manufacturing departments will also be involved. For large-scale investments, top management will also be involved.

## 9.8  Time travel exercise method

For the development stage, a new team is composed to execute the development stage. One of the challenges in this stage is that the new team feels they are the owners of the new project. To establish ownership, it is not enough to transfer the existing knowledge of previous stages. It requires intensive dialogue to find out what the project is about. This time travel exercise is a method to go back to the feasibility stage where the commercial-scale and the down-scale pilot plant were designed, and to repeat its main steps in a pressure-cooker process. The following steps have to be repeated:

- SDGs
- PI design method
- VIB analysis
- Multidisciplinary assessment.

The following procedure can be used for SDGs:
1) The development team discusses the importance of the first SDG for the project, in short (see Chapter 3).
2) The discussion leader presents the final judgment of the feasibility team with respect to the relevant SDGs.
3) The discussion is closed by a short evaluation of the differences between steps 1 and 2.
4) The steps 1 to 3 are repeated for every SDG.

The following procedure can be used for the PI design method:
1) One of the design engineers of the feasibility stage presents a short introduction to PI (see Chapter 2) and to the present process design.
2) The discussion leader invites the members of the team to reflect on the question of how the present process can be improved by the application of the different concepts of PI.
3) One of the developers of the concept stage tells the story starting from the discovery stage up to the feasibility stag: what experiments have been done, what considerations played a role, what dilemmas cropped up, and what decisions were made.

The following procedure can be used for the VIB analysis:
1) One of the engineers presents a short introduction to the VIB method (see Chapter 4).
2) The discussion leader invites the team to reflect on the values of engineers.
3) The discussion leader presents the conclusion of the feasibility stage with respect to the values of engineers.
4) The discussion is closed with an evaluation of the differences between steps 2 and 3.
5) The discussion leader invites the team to reflect on the interests of stakeholders.
6) The discussion leader presents the conclusion of the feasibility stage with respect to the interests of stakeholders.
7) The discussion is closed with an evaluation of the differences between steps 5 and 6.
8) The discussion leader invites the team to reflect on the beliefs in society for the relevant allocation.
9) The discussion leader presents the conclusion of the feasibility stage with respect to the beliefs in society for the relevant allocation.

10) The discussion is closed with an evaluation of the differences between steps 8 and 9.

The following procedure can be used for the multidisciplinary assessment:
1) One of the engineers presents a summary of the multidisciplinary assessment of the feasibility phase.
2) The topics on which the members of the multidisciplinary assessment agree are discussed, in short. Particular attention has to be given to the underlying arguments.
3) The topics on which the members of the multidisciplinary assessment disagree are discussed. In particular, attention has to be given to the arguments of the different disciplines and the decision made in the gate evaluation.

## 9.9 Multidisciplinary evaluation method

In Chapter 5 of this book, we have discussed the multidisciplinary assessment method. The key purposes of the method are twofold: first, to identify and highlight the specific knowledge of different departments and/or different hierarchical levels of the organization; second, to integrate the different bodies of knowledge. In this phase, the knowledge of the practices of R&D, OPS, and M&S have to be assessed. Also, the practices of SHE department have to be assessed. It is essential that the results of the foregoing steps (from scoping to front end engineering design) are discussed with the different disciplines in separate meetings. The main reason is to make the contribution of every discipline explicit and sharp. After that, the different contributions can be discussed and integrated in a joint meeting.

A multidisciplinary assessment for this phase can be done in the following way:
1) The project team discusses the results of the development stage with a broad group of engineers of the R&D department. This meeting focuses on the technical feasibility of this project. The capability of the process to meet the specifications (technical, economic, SDG, VIB, and so) have to be discussed and evaluated. Additionally, the scaling-up risks are discussed.
2) The project team discusses the results of the development stage with a broad group of managers and process engineers of the OPS department. This meeting focuses on the functioning of the pilot plant. Basically, it is all about process control and all human activities that lead to process control. Also, the proposal for the final design is discussed extensively.
3) The project team discusses the results of the development stage with a broad group of colleagues of the M&S department. In this meeting, the market opportunities for the intensified process are discussed in great detail. In particular, the expected added value for the various customers is discussed.

4) The project team discusses the results of the development stage with a broad group of colleagues of the SHE department.

5) The project team analyses the discussions with R&D, OPS, M&S, and SHE departments. Similarities in views and differences in opinion have to be identified. Attention has to be paid especially to differences of opinion. A concept multidisciplinary assessment report is written.

6) The project team invites two or three colleagues each, from the R&D, OPS, M&S, and SHE departments to discuss the concept multidisciplinary assessment report. The importance and the impact of differences in opinion have to be discussed. The discussions about these differences in opinion should preferably lead to (a) agreement about the issue, (b) the agreement to disagree about the issue, or (c) the recognition that more research is required to judge the issue.

7) Based on this discussion, the project team finalizes the multi-disciplinary assessment report. It has to be noted that differences of opinion that could have an influence on the gate evaluation have to be made explicit in the report.

## References

[1]    Harmsen, J, et.al., Product and process design driving innovation, De Gruyter, Berlin, 2018.
[2]    Harmsen, J, Industrial process scale-up – A practical innovation guide to idea to commercial implementation, 2nd revised edition, Elsevier, Amsterdam, 2019.
[3]    Harmsen, GJ, Reactive distillation : The frontrunner of Industrial Process Intensification: A full review of commercial applications, research, scale-up, design and operation, Chemical Engineering & Processing: Process Intensification, 2007, 46(9), 2007, 774–780.
[4]    Schwalbe, T, Microstructured reactor systems, Chimia, 2002, 56(11), 636–646.
[5]    Vogel, GH, Process development – from the initial idea to the chemical production plant, Wiley-VCH, Weinheim, 2005.
[6]    Bakker HLM, and de Klein, JP, ed. Management of engineering projects – people are Key, NAP, Nijkerk, 2014.

# 10 Engineering procurement and construction (EPC) stage

## 10.1 Introduction

The engineering procurement and construction (EPC) stage is also called the Execution stage. But the latter includes start-up. Here, we separate the EPC from the start-up as the EPC stage involves people different from those in the start-up stage and with different roles. In this separate chapter on the EPC stage, the focus is on specific critical aspects in relation to radical process innovation, using process-intensification methods and technologies.

The purpose of the EPC stage is to deliver the commercial-scale process to the manufacturer. The steps and supporting methods for this stage are shown in the Figure 10.1. The EPC contractor, in close cooperation with his client, executes this stage.

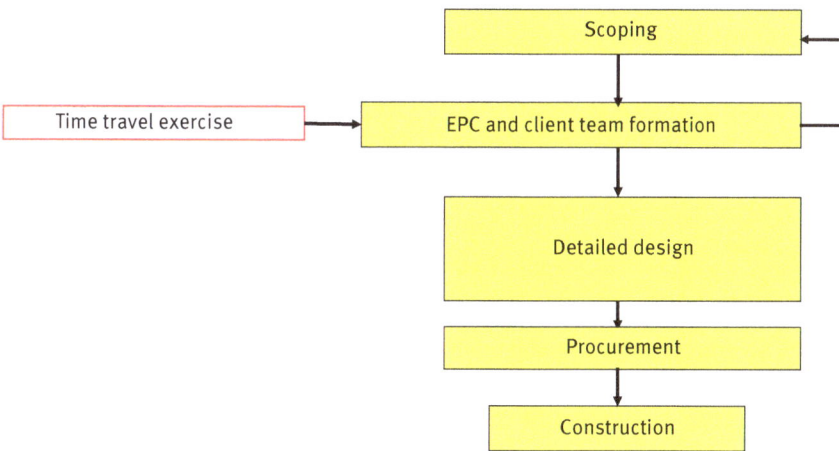

Figure 10.1: Engineering procurement and construction: steps and methods.

Section 10.2 describes the steps of this stage. Section 10.3 describes the supporting method of the steps, indicated in Figure 10.1.

https://doi.org/10.1515/9783110657357-010

## 10.2 EPC project steps

### 10.2.1 Scoping

In the scoping step, the goal and requirements of the commercial-scale process are defined. When the process is first of its kind, as is the case in all innovation projects, then that process needs additional precaution measures to reduce the risks to an acceptable level. The risks concern the technical reliability, the operational reliability, and a sufficient market for the product. A scope statement concerning risk reduction needs to be in place. An important aspect in this respect is the choice of the production capacity.

For oil and gas branch of a totally new large process complex, such as the gas-to-liquids plant of Shell, the desired commercial-scale capacity involves an investment of 10–20 billion Euros and the company then, often, decides to go on a so-called demonstration process scale, first. The production capacity is then typically a factor 10–50 lower, and the investment is typically a factor 10 lower. For processes containing only a few steps such as for reactive distillation with an investment of typically 10–100 million Euros, a full-scale plant is chosen and not a demonstration plant scale. Harmsen provides detailed information of the Shell GTL demonstration plant [1].

For all other branches the choice of the process capacity is, in most cases, strongly dependent on the product market uncertainties. Often, a small scale is chosen to scout the market. After a few years, when the market appears to be there, a larger scale process is designed. For bulk chemicals, where markets grow at a slow steady state, a demonstration plant is hardly selected. In nearly all cases, a full commercial scale is directly chosen [1]. Examples of such choices are provided in Chapters 14 and 15.

### 10.2.2 EPC contractor selection and team formation

The EPC contractor selection is always a very important part of the EPC stage. This company designs the process with enormous detail. Every piece of equipment, connecting pipes, and electrical connection has to be defined in size, shape, and type of material. This is the engineering part of the stage. The contractor should be experienced in the industrial branch where the new process is envisaged. A contractor experienced in food process design is not suited for a bulk chemicals process design, and vice versa. Bakker emphasizes that the choice of the EPC contractor should be made by the project leader of the manufacturing company and not by its purchasing department. The latter is likely to turn into a catastrophe [2].

The contractor or subcontractor should also have experience in engineering the intensified process with its special equipment. For a new process involving an intensified technology, often, a specific technology provider will be consulted for the detailed design. In the procurement step, the PI technology will then be bought.

The contractor works in close cooperation with his client to ensure that in all detail, the new process fulfills its intended purpose. During the EPC stage, additional questions need answers and additional decisions have to be jointly taken. Understanding each other and the task at hand become vital. A special organizational project structure is to be set up. Bakker's book on Project management describes the organizational structure in detail [2]. Harmsen provides some important team success factors [3]:

– Leadership
– Trust
– Team building
– Risk identification, monitoring and appropriate measures

To create trust, teambuilding, knowledge transfer, and risk identification, a special method, called Time travel exercise, is provided for this joint team. The method is provided in Section 10.3.

### 10.2.3 Detailed design

Detailed design is an enormous effort with many specialists involved in each detailed design aspect. Because of this, a large project organization will be set up. It is beyond the scope of this book to describe, in some detail, what is involved in the execution of the detailed design. Bakker provides detailed information for this project step [2].

For process intensified processes having less equipment, needing fewer valves and pumps and controls and thereby less connections, the detailed design will be a smaller effort. This reduces risks of failure and reduces detailed design cost. However, strong involvement with the PI technology provider in this project step will be needed to avoid misunderstandings on the detailed design that may lead to a failure at the start-up.

### 10.2.4 Procurement

All major equipment is procured by the contractor. The reason for having a contractor who has experience in the specific industry branch is particularly relevant for the procurement. Such a contractor knows the equipment suppliers for this branch and knows which suppliers have a good track record. A contractor only experienced in a different branch could easily select the wrong suppliers who provide equipment unsuitable for the target branch. Harmsen provides several examples of wrong selection of an EPC contractor and wrong procurement of major equipment, resulting in long start-ups [3].

Procurement of the process-intensified section of the new process is even more critical to the success. There are many experienced providers of PI processes, but there are also many new start-up companies providing PI processes. The latter are often very good in providing the specific equipment, but may be not so good in providing the integral process design. The combination of an experienced EPC and a technology provider with a well-designed down-scaled pilot plant can work well, as shown by the successful commercial-scale PI processes described in Chapter 2. Chapter 13 shows an example of the intimate cooperation between the PI technology provider and the EPC contractor.

### 10.2.5 Construction

In the construction step, the process is erected. This construction can be on site. It can also be mainly carried out at the EPC location. In that case, the whole process is constructed in such a way that it can be transported to the manufacturing site. This is called skid-mounted. The process is constructed in one or more skids: frames to which the process is connected. This ensures safe transport. The commissioning of the process is then mostly carried out at the EPC location. Commissioning involves cleaning the process by flushing with water and nitrogen, testing instrumentation and control systems and remedying detected errors.

The construction ends with commissioning the process. This means that all process equipment, including pipes are cleaned and all instrumentation and controls are tested. At the end of the commissioning, the process is handed over from the project management to the process operation management. This handover is described in the start-up chapter.

## 10.3 Time travel exercise method

After the EPC Contractor selection, a new team will be established (See 9.2.2.). This will be a team of EPC engineers and at least one project engineer of the client company. Other engineers from the client operation department may also be involved.

It is of utmost importance to create trust between its members, to share all relevant knowledge, and to identify risks. The objective to the time travel exercise method is to realize all these goals. By sharing information, by having dialogues about the key issues of the project, and by setting off together, trust will develop, and a good team spirit will grow. It goes without saying, that for the time travel exercise, an adequate setting has to be chosen. In addition, the program has to be designed in such a way that team building will grow.

The key of the time travel exercise is twofold: (1) to go back to the feasibility stage (SDGs, VIB) and to the development stage (PI design method, multidisciplinary assessment) and to repeat its main steps in a pressure-cooker process, (2) to go back to the history of the EPC contractor to understand why their employees work as they work, how they control quality, and how they identify and manage risks. It goes without saying that the time travel exercise is not only about knowledge transfer but also about speaking the same language and understanding each other's professional culture.

The following steps have to be done by the whole team in a pressure-cooking process:

- SDG's
- PI design method
- VIB analysis
- Multidisciplinary assessment
- Contractor's story (part of multidisciplinary assessment).

The following procedure can be used for SDGs:
1)  The team discusses the importance of the first SDG for the EPC stage, in short (see Chapter 3).
2)  The discussion leader presents the final judgment of the feasibility team with respect to the relevant SDG.
3)  The discussion is closed by a short evaluation of the differences between steps1 and 2.
4)  The steps 1 to 3 are repeated for every SDG.

The following procedure can be used for the PI design method:
1)  One of the engineers of the R&D departments presents a short introduction to PI, its opportunities, and its domains (see Chapter 2).
2)  One of the engineers of the R&D department tells the story of the relevant PI project starting from the discovery stage up to the development stage –including which experiments have been done, what considerations played a role, what dilemmas cropped up, and what decisions were made.
3)  The discussion leader invites the members of the team to reflect on these presentations and to identify possible critical points and possible risks.

The following procedure can be used for the VIB analysis:
1)  One of the engineers presents a short introduction to the VIB method (see Chapter 4).
2)  The discussion leader invites the team to reflect on the values of engineers.
3)  The discussion leader presents the conclusion of the feasibility stage with respect to the values of engineers.

4)  The discussion is closed with an evaluation of the differences between steps 2 and 3.
5)  The discussion leader invites the team to reflect on the interests of stakeholders.
6)  The discussion leader presents the conclusion of the feasibility stage with respect to the interests of stakeholders.
7)  The discussion is closed with an evaluation of the differences between steps 5 and 6.
8)  The discussion leader invites the team to reflect on the beliefs in society for the relevant allocation.
9)  The discussion leader presents the conclusion of the feasibility stage with respect to the beliefs in society for the relevant allocation.
10) The discussion is closed with an evaluation of the differences between steps 8 and 9.

The following procedure can be used for the multidisciplinary assessment:
1)  One of the manufacturing engineers presents a summary of the multidisciplinary assessment of the development phase.
2)  The topics on which the members of the multidisciplinary assessment agree are quickly discussed. In particular, attention has to be given to the underlying arguments.
3)  The topics on which the members of the multidisciplinary assessment disagree are discussed. Specifically, attention has to be given to the arguments of the different disciplines and the decision made in the gate evaluation.
4)  The EPC contractor project leader is invited to tell his story. The following points need to be addressed and discussed in the team: the history of his company, the construction philosophy, the quality policy, risk assessment, and risk management.

## References

[1]   Harmsen, J, Industrial process scale-up – A practical innovation guide to idea to commercial implementation, 2nd revised edition, Elsevier, Amsterdam, 2019.
[2]   Bakker HLM, de Klein, JP, ed. Management of engineering projects – people are key, NAP, Nijkerk, 2014.
[3]   Harmsen, J, et.al. Product and process design driving innovation, De Gruyter, Berlin, 2018.

# 11 Start-up and normal operation stage

## 11.1 Introduction

It is important to mark the start-up of a new process as separate from normal operation. I have witnessed twice a start-up where the start-up was seen as part of normal operation. No start-up team was formed, and no start-up plan was made. In fact, no one in the local operation organization even had the notion that start-up of the process was new and needed special start-up measures. In both cases, the start-up went horribly wrong. After half a year of trouble to start-up and operate the processes, outsiders from local operation were involved. Even analyzing the problems was hard. No formal reporting of the start-up was available.

It is worth noting that, according to the Merrow definition, a process is new if there is no process in commercial operation for that particular feedstock, product, or catalyst. Often when the individual process unit operations are conventional it is assumed that the process is conventional, and no start-up plan is needed. In reality, such a process is new, and a start-up plan is needed. If the process is being applied in other companies, but not in this particular company, then it is worthwhile to walk through the process development in the past in the other companies and their commercial-scale operation so that the process becomes familiar.

Figure 11.1 shows the start-up project steps to be taken and their supporting methods. Section 11.2 provides information of the steps, in general. Sections 11.3–11.5 provide methods that support the project steps.

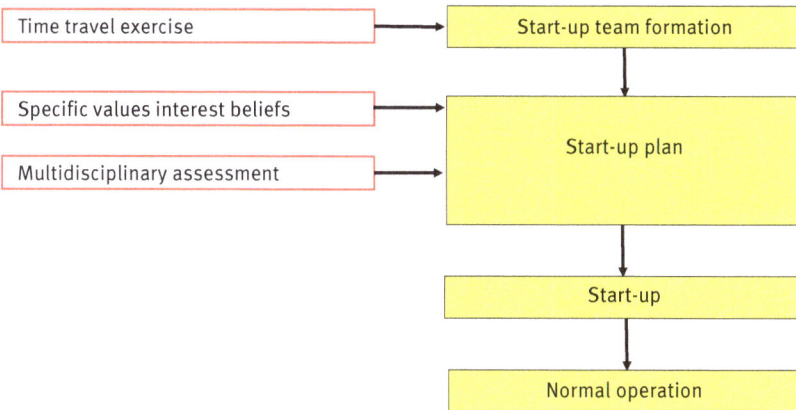

**Figure 11.1:** Start-up stage: steps and methods.

https://doi.org/10.1515/9783110657357-011

## 11.2 Project steps

### 11.2.1 Start-up team formation

Start-up of a new process will be done by the process operators. It is advised to have a leader of the operators, a shift supervisor who has experience with other start-ups of novel processes. If the process operations leader does not have this experience, then he/she will need an advisor with this experience.

It is also advised to have a start-up support team to provide guidance to the operators. This start-up support team formation is an important element of the start-up preparation. For process-intensified start-up this is even more important due to the additional process novelty of unconventional process steps. Harmsen provides guidelines for this start-up support team formation and for additional supporting experts, which can be called upon when needed [1]. This support team will have an experienced process operator, a process engineer, and a process control specialist. A process developer and a process chemist involved in the concept and development stages of the project should be on stand-by for questions.

A special method is provided in Section 10.3 to build the knowledge of this start-up support team.

### 11.2.2 Start-up plan

A well thought-out start-up plan should be made to ensure that all knowledge of previous innovation steps is on board. Also, this should ensure that risk items are noted down and that special precaution measures are taken to counteract those risks. Harmsen [1] provides the following critical success factors for a start-up:
– Potential problem analysis and precaution measures taken
– Complete start-up support team formed
– Operators trained for start-up and normal operation
– Complete start-up plan made
– Documentation of start-up done

If these factors are executed, then the start-up time can be a factor 10 shorter than the industry-averaged start-up time defined by the Merrow correlation under the premise that the critical success factors in the development and design stages are executed.

The development and design critical success factors are [1]:
– Process recognized as new
– Integrated down-scaled pilot plant available
– R&D and EPC knowledge integrated
– Scale-up knowledge for unit operations are available.

Section 11.3 provides a time travel exercise method to familiarize the start-up team with these critical success factors for the process to be started up, and in this way to support the start-up plan. Section 11.4 provides a method to bring values, interests, and beliefs into the start-up picture, so that the start-up members are aware of the importance of these items. Section 11.5 supports the assessment of the start-up plan by a multidiscipline team, so that flaws are identified and remedied.

### 11.2.3 Start-up

The start-up is carried out by operators of the manufacturing company, supported by the start-up team. The actual start-up starts when feeds enter the process for the first time.

Prior to the start-up, commissioning of the process has been carried out. This is, in general, a co-operation between the EPC contractor and the manufacturing company. A pre-start-up audit marks the end of the commissioning. In the oil and gas branch, a statement of fitness is then prepared. It is a handover document signed by the project manager and the future process owner to confirm that all safety requirements have been addressed [2].

Reporting the start-up is important for several reasons. If later, during normal operation a problem occurs, then a check on what happened during the start-up can help in analyzing the cause of the problem. For the start-up of a similar process, learning points from the start-up can be considered. If all start-ups are reported, then generic lessons can be drawn from start-ups and put into a practice report on start-ups.

### 11.2.4 Normal operation

Information about normal operation of the new plant is very interesting for the next design of the process. Often a first-of-its-kind process is designed with redundancies. In normal operations, these redundancies become manifest. The simplest redundancy may show up as a larger than design production capacity. Reporting the operational results after one or two years in a review report will be very useful if a second design is to be made using the same process. Using the review report, the second design can often be 40% lower in investment cost than the first design. Further guidelines for normal operation of the new plant are beyond the scope of this book.

## 11.3 Time travel exercise method

The time travel exercise method is well suited to create trust between the start-up leader and the start-up support team members, to share all relevant knowledge, and to identify risks in the start-up phase. By sharing information, by having dialogues about the key issues of the project, and by setting off together, trust will develop, and a good team spirit will grow. For the time travel exercise, an adequate setting has to be chosen – preferably in a conference center – so that the entire team can work in a concentrated way and there is a lot of time to get to know each other informally.

The key purposes of the time travel exercise are threefold. First, to understand the intensified process by going back to the previous development stages and to repeat its main steps in a pressure cooker process; secondly, to understand the process final design, construction and installation by going back to the Engineering procurement and construction phase (EPC); and finally, to develop a joint understanding of the critical items of the start-up based on the history of the development of the intensified process.

Generally, a time travel exercise is a substantial investment in the team. Such an investment pays for itself quickly because the quality and efficiency of the collaboration greatly increases. It also leads to a decrease in the risks of the start-up.

The following steps have to be done by the whole team in a pressure-cooking process:
–  SDGs
–  PI design method
–  Multidisciplinary assessment
–  EPC story

The following procedure can be used for SDGs:
1) The team discusses shortly the importance of the first SDG for the start-up stage (see Chapter 3).
2) The discussion leader presents the final judgment of the feasibility team with respect to the relevant SDG.
3) The discussion is closed by a short evaluation of the differences between steps 1 and 2.
4) Steps 1–3 are repeated for every SDG.

The following procedure can be used for the PI design method:
1) One of the engineers of the R&D departments presents a short introduction to PI, its opportunities, and its domains (see Chapter 2).
2) One of the engineers of the R&D department tells the story of the relevant PI project, starting from the discovery stage up to the development stage, including which experiments have been done, what considerations played a role, what dilemmas cropped up, and which decisions were made.
3) The discussion leader invites the members of the team to reflect on these presentations and to identify possible critical points and possible risks.

The following procedure can be used for the multidisciplinary assessment:
1)  One of the engineers presents a summary of the multidisciplinary assessment of the development phase.
2)  The topics on which the members of the multidisciplinary assessment agree are quickly discussed. Attention has to be given, especially, to the underlying arguments.
3)  The topics on which the members of the multidisciplinary assessment disagree are discussed. In particular, attention has to be given to the arguments of the different disciplines and the decision made in the gate evaluation.

The following procedure can be used for the EPC story:
1)  One of the engineers of the contractor tells the story of the engineering procurement and construction of the intensified process.
2)  The risk analysis performed in the EPC stage is repeated with all team members. The objective is not only to transfer knowledge to the new members of the team but also to integrate as well as possible, the local knowledge about production. This risk analysis is one of the inputs for the start-up plan.
3)  The multidisciplinary evaluation at the end of the EPC stage is presented. Attention is particularly given to critical decisions and possible risks. This evaluation is also one of the inputs for the start-up plan.

## 11.4  Specific values, interests, and belief methods

In the feasibility stage, an extensive VIB analysis was done. The analysis has to be repeated in the start-up phase. The main reason is that since that time, a lot of new information has come to light. Additionally, the VIB analysis can be done very specifically in view of the chosen allocation.
The following procedure can be used for the VIB analysis:
1)  One of the engineers presents a short introduction to the VIB method (see Chapter 4).
2)  The discussion leader invites the team to reflect on the values of engineers.
3)  The discussion leader presents the conclusion of the feasibility stage with respect to the values of engineers.
4)  The discussion is closed with an evaluation of the differences between steps 2 and 3.
5)  The discussion leader invites the team to reflect on the interests of stakeholders.
6)  The discussion leader presents the conclusion of the feasibility stage with respect to the interests of stakeholders.

7) The discussion is closed with an evaluation of the differences between steps 5 and 6.
8) The discussion leader invites the team to reflect on the beliefs in society for the relevant allocation.
9) The discussion leader presents the conclusion of the feasibility stage with respect to the beliefs in society for the relevant allocation.
10) The discussion is closed with an evaluation of the differences between steps 8 and 9.

It should be noted that enough time is needed to do this exercise. It could easily take a day.

## 11.5 Multidisciplinary assessment method

In Chapter 5 of this book, we have discussed the multidisciplinary assessment method. The key purpose of the method in this stage is to assess and improve the start-up plan from different perspectives. It is something like a final check. In fact, everyone thinks it superfluous because the start-up plan has been put together with great care. And yet nobody wants to run the risk that something important has been overlooked.

The multidisciplinary assessment can be done in the following way:
1) Multidisciplinary evaluation by the contractor. Once more, the contractor is invited to do an assessment with the members in the start-up team and with their relevant specialists. The main objective is to evaluate the start-up plan. A report is made.
2) Multidisciplinary evaluation by the R&D department of the company. The assessment is done with the members of the team, their relevant specialists, and relevant managers. The main objective is to evaluate the start-up plan. A report is made.
3) Multidisciplinary evaluation by the engineers who will run the equipment in the operation stage. The main objective is to evaluate the start-up plan from the view of the start-up and the subsequent operation stage. A report is made.
4) Representatives of the contractor, the R&D department of the company, and the engineers and management of the plant will discuss the different reports. Attention will be given especially to differences of opinion, if any, about critical aspects of the design, possible risks during start-up, and so on.
5) In case big differences of opinion or unexpected large risks appear, the project leader organizes a joint meeting with the relevant parties, managers, and engineers.

It should be noted that in steps 1–3, members of the different parties that are not a part of the start-up support team are invited. The reason is that new people who can evaluate the start-up plan with a fresh mind have to be brought in.

## References

[1]   Harmsen, J, Industrial process scale-up – A practical innovation guide to idea to commercial implementation, 2nd revised edition, Elsevier, Amsterdam, 2019.
[2]   Bakker HLM, de Klein, JP, ed. Management of engineering projects – people are Key, NAP, Nijkerk, 2014.

Part C: **Industrial practice cases**

# 12 Summary PI industrial practices

## 12.1 Introduction

The process-intensified (PI) industrial cases described in Chapters 13–17 of this book have been obtained from a variety of industry branches – food, pharma, biomass conversion, and bulk chemicals.

We are trying to see what these very different cases have in common with regard to the role of innovation practices as described in Chapters 2–5. We have also made an effort to see other common aspects in these cases that can be of benefit to designers and developers of future innovation projects. We found that the role of education and universities played an important role in most cases, and therefore added that topic.

## 12.2 PI domains and design

They cover all PI categories, plant-wide space, function integration, temporal, thermodynamic, and equipment space. Hence, it is clear that PI design is applicable to all process industries.

In several cases, the design starts with plant wide, followed by function integration. This is clearly a logical and fruitful sequence resulting in breakthrough novel designs. Smart heat integration, without applying heat exchangers, is also often applied.

## 12.3 Sustainable development goals

Most innovation cases were executed prior to the introduction of the sustainable development goals (1987), the exception being the BTG-BTL case, where the SDGs played an explicit role. Improvements in health, safety, environmental, economics, and social aspects; hence, improvements in planet, prosperity, and people dimensions played a role in all cases. So implicitly, SDGs played a role in all cases.

## 12.4 Values of engineers

Values of engineers played an overriding role in the Eastman, Molatec, and BTG-BTL cases, where they drove the idea to successful implementation against the odds. The common aspect in this respect seems to be that the engineers wanted to work on something new and far better than conventional technologies. This shared

https://doi.org/10.1515/9783110657357-012

value also played an important role in the Shell OMEGA development, where engineers of Mitsubishi shared the same passion to make the new technology a success.

## 12.5  Interest of stakeholders

In all cases, the interest of stakeholders played an important role. In the BTG-BTL case, the contract with the Friesland-Campina company in Borculo to promise to buy the bio-oil ensured that investment capital from the other stakeholders could be obtained. In the pharma case, the regulator institutes play a major role ensuring that the product (medicine) quality is maintained in commercial-scale production.

## 12.6  Beliefs of local society

It is to be noted that all innovation cases had a strong interaction with the local society. Eastman developed and implemented the process in Kingston Tennessee, well aware of their responsibility for the local city community being the main employer. Therefore, they wanted a very competitive process for methyl acetate, so that they were robust enough to face outside competition. Mitsubishi developed the glycol process without being a market player of technology in that field. In the end, the process was implemented first three times in East Asia, because it was new and more efficient than conventional processes; and one of the manufacturers wanted this Mitsubishi process technology from Shell, the market leader in ethylene glycol process technology licensing. BTG-BTL invented, developed, designed and implemented their process within a radius of 30 km (Twente region) with only local parties involved; the Twente culture of mutual trust played an important role. Cyclic distillation was developed and implemented in Ukraine. It seems that being part of a local community played an important role in pursuing these breakthrough innovations.

## 12.7  Knowledge transfer and multidisciplinary co-operation

Knowledge transfer over the stage boundaries and multidisciplinary knowledge exchange and evaluations played an important role in all described cases. The BTG-BTL case is exemplary in this respect. Information, experience, and knowledge, particularly, tacit knowledge was carried among persons from research to development, to design, and to implementations. Multidisciplinary evaluations were carried out by one person, the director.

Also, in the Eastman case, knowledge transfer by transferring people played an important role. In that case, close proximity of engineers and management (across the corridor) helped Eastman enormously in building trust for rapid decisions.

Multidisciplinary co-operation between R&D, Operations, and Marketing & Sales played an important role in all cases, but no formal organizational structure with formal stage-gate evaluations appeared to be in place in the companies. Co-operation between these disciplines seems to be part of the company cultures, rather than forced by an organization structure.

## 12.8  Role of universities and education

University education of the development engineers in generating the novel design played a very important role for the Eastman, Molatec, and BTG-BTL cases. Their new concepts were conceived due to university education. Advanced modeling played an important role in the Molatec, Eastman and OMEGA cases, but played a minor role in the BTG-BTL and pharma cases.

## 12.9  Epilogue

The five cases are a too small sample for scientific proof of the industrial innovation practice theories provided in Chapters 3–5. However, the cases clearly show that these soft aspects played an important role in the successful innovation, from idea to commercial-scale implementation.

In engineering, it is perfectly okay to use a theory that already worked in some cases, if applying that theory is plausible and has benefits. Scientific proof follows later, by using the successful cases. Or as the saying goes, "Science owes more to the engine, than the engine owes to science." A prime example is the steam engine that led to the second law of thermodynamics.

# 13 Rotating cone reactor biomass pyrolysis process

## 13.1 Background information BTG-BTL process

The BTG-BTL pyrolysis process converts biomass feedstock into bio-oil as the main product. The other product is high pressure steam which is used to dry the biomass feed. Excess steam is used to generate electricity and for district heating.

The core technology is a reaction system consisting of a rotating cone reactor (RCR). In 2 s, it converts the biomass to oil, gas, and char. And a fluid bed reactor called combustor, combusting char, and noncondensables. Hot sand of that combustor is recycled back to the RCR.

Figure 13.1 shows the major elements of the final process design. The development leading to this final design and implementation is described in detail in the next sections.

## 13.2 Discovery stage

Michiel Groeneveld and Wim van Swaaij of University Twente are the original inventors of the RCR for fast pyrolysis. They invented it at the end of the 1980s. The reactor concept is based on the idea that no inert gases are required for biomass heating for pyrolysis, instead rapid mixing of biomass with hot sand is a better way. This rapid mixing is enhanced by mechanical mixing of biomass and hot sand by a rotating cone. Wolter Prins was instrumental in this development, first as a supervisor of his first PhD student, Bert Wagenaar. The original concept was to supply heat from the outside of the cone. However, initial tests showed that this was not appropriate and the concept of using sand as a heat carrier was introduced [1].

Many others contributed to the original idea with less or more success. However, Wagenaar, who published in 1994 his thesis on the modeling (reactor modeling, heat transfer) and experimental results of the laboratory-scale RCR concept, was instrumental in the pyrolysis of wood. The system was operated batchwise [1,2].

## 13.3 Concept stage

An innovative integrated process design was thereafter made, in the PhD work of Arthur Jansse. It consisted of two interconnected fluid-beds. In one bed, the cone was placed to produce bio oil and the other was used for generating heat by combustion. Sand was recycled from the combustor to the pyrolysis reactor by a riser.

https://doi.org/10.1515/9783110657357-013

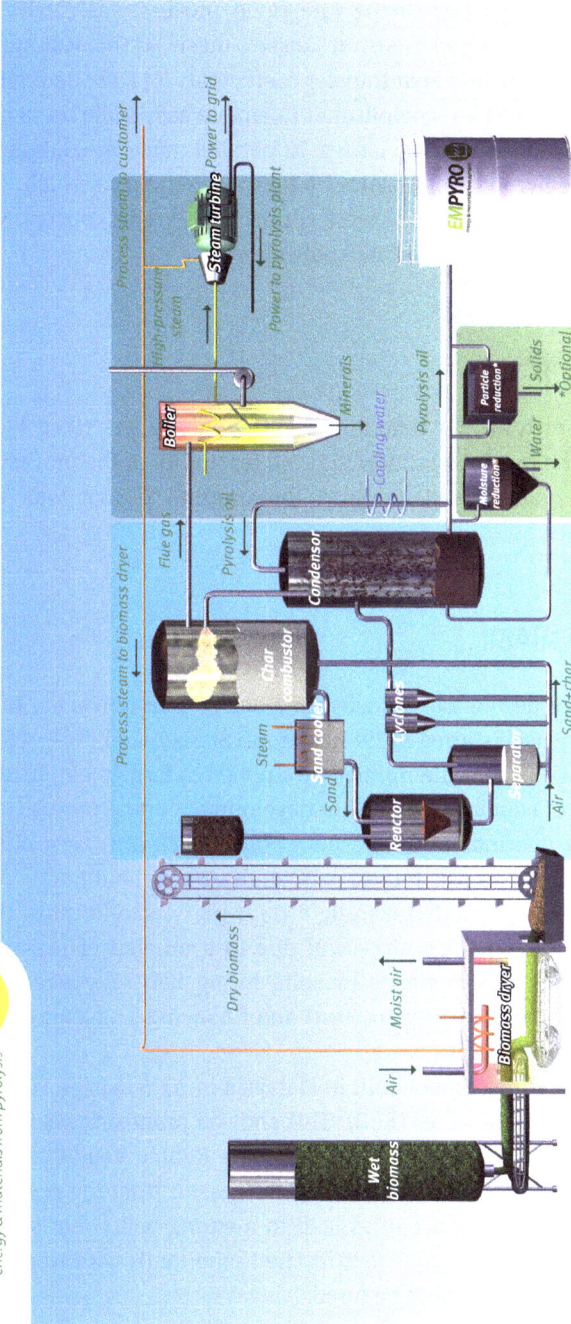

**Figure 13.1:** BTG-BTL Process.

[3]. Experimental proof of concept of the integrated process was carried out by Robbie Venderbosch and published in Arthur Jansse's thesis [4] that was supervised by Prins. Combustion kinetics and heat transfer coefficients [6] were determined [5].

However, the concept was too complicated to enable easy scale up to commercial scale, mainly due to pressure drop issues. In parallel, another continuous unit was constructed for the pyrolysis of plastics by Ron Westerhout in which, both pyrolysis reactor and char riser combustor were connected [7]. This concept was considered to be suitable for commercial-scale applications.

## 13.4 Feasibility stage

This stage was not known at that time and so little can be described here. It is clear from the text in the previous section that the feasibility of the process design for commercial scale was considered and the original design was adapted accordingly.

## 13.5 Development stage

The Westerhout concept design was transferred from Twente university to BTG in 1994. The BTG company itself started in 1987 by Huub Stassen and Roland Siemons, as a spin-off from the UT. The scale-up method applied is largely empirical in nature and required a lot of stamina. The further development of the process took several years with a lot of information from a pilot plant of 180–250 kg/h of biomass that was started up in 1998 and which used parts of the Westerhout unit.

In parallel, activities were undertaken by a third party who licensed the technology from BTG, but this was not successful due to a number of reasons, chief amongst which were lack of experience (despite hiring Bert Wagenaar as a key actor), management failures (too large too fast) and financing problems (leading to bankruptcy) on the side of the third party.

A first demonstration process was built in Malaysia in 2005 using a very complicated feedstock, empty fruit bunches (EFB). This showed promising results in relation to the process design. Key actors in this unit were Robbie Venderbosch, Daan Assink, and Elwin Gansekoele. Though not very successful from the point of view of a commercial operation, these activities lead to a strong belief that such a unit could be operational on a commercial basis in The Netherlands with wood as feedstock and if adequate personnel could be hired.

Parallel to this development, a small pyrolysis test facility of 5 kg/h was built. It was instrumental in resolving scale-up issues. It also enabled tests with more challenging feedstocks. BTG has successfully tested over 45 different kinds of feedstock over the years, ranging from relatively simple feedstocks like wood to difficult feedstocks like sludge or EFB. Evert Leijenhorst and William Wolters were, in particular, involved in this testing.

The specific success of the Malaysian unit led to interest from Gerhard Muggen and Ardy Toussaint to be involved (also financially) in a further scale-up and commercialization of fast pyrolysis, based on the retaining cone reactor concept. This led to the project "Empyro," a European project based on poly generation of various products and set-up by, amongst others, Bert van de Beld and Robbie Venderbosch, and granted by the European Commission in 2009. The project was managed by a.o. Patrick Reumerman with strong involvement of Muggen and Toussaint in the further technical development. This eventually led to the founding of the *company* Empyro B.V. in 2014.

## 13.6 Engineering procurement and construction stage

The envisaged commercial-scale plant was for EMPYRO BV. Empyro BV was at that time jointly owned by BTG BioLiquids, *Tree Power*, the province of Overijssel, and a private investor. They jointly invested 10–20 M€ (approximately). A critical reason for providing the investment was a contract signed with Friesland Campina, Borculo, The Netherlands, to buy the biopyrolysis oil produced.

The EPC stage for the EMPYRO process started in March 2013 with a contract with the EPC companies ZETON, Stork-Thermeq, HoSt, and BTG-BTL. Gansekoele of BTG-BTL, involved in all development, was transferred to ZETON to ensure maximum knowledge transfer to the EPC stage. In the detailed engineering, a strong focus was placed on all critical aspects of the mechanical engineering for the RCR and other process elements. All learning points of the pilot plant and the Malaysia plant were taken into account.

A contract was signed for a commercial-scale plant for the conversion of wood residues into pyrolysis oil, process steam and electricity at EMPYRO Hengelo (O). The plant location is the industrial site of AKZONOBEL Hengelo (O). Construction at that site started in January 2014.

The company ZETON was selected as the prime EPC contractor. This company has large experience in constructing modular pilot plants and small-scale commercial plants and had also executed the EPC stage for the Malaysian BTG-BTL plant. The company is located at Enschede, the same city as BTG-BTL.

The commercial-scale design of the EMPYRO plant consists of 22 modules. These modules were constructed and dry tested at the ZETON location. Then, the skid-mounted modules were disconnected and transported to the production location at Hengelo (O).

The modular design means that the detailed design can be re-used for follow-on processes. Dry-testing at the construction site of ZETON means that start-up at the production site is rapid, requiring little extra operators.

## 13.7 Start-up stage commercial-scale plants

The first commercial-scale process, in fact a demo-plant (see also Section 13.5), was a 2 t/h pyrolysis plant; designed, constructed, and delivered to Malaysia in 2005. In the factory – located close to an existing palm mill –EFB is converted into pyrolysis oil. Wet EFB from the palm is converted into approximately 1.2 t/h pyrolysis oil.

The second commercial-scale process of 5 t/h, with wood biomass as feed, was for EMPYRO, Hengelo (O). Start-up of the installation commenced in early 2015 and production was gradually increased. In October 2017, the Empyro plant reached 100% of its nameplate capacity. Empyro passed then the 20 million liter mark, making it the world leader in pyrolysis oil production. The time from the start to this nameplate capacity was 21 months [8].

This start-up time is compared to the start-up time of new industrial plants according to the Merrow correlation to see whether a new process for biomass feed would be in accordance with the correlation. For this correlation to be determined, some parameters have to be known. The first parameter is the number of new process sections ($N$). The EMPYRO process sections are: Biomass drying, Reaction, Flue gas treatment, and Product work-up. The EMPYRO process feedstock being wood, a lignocellulose, it is different from the Malaysian plant which had its feedstock as EFB which is not a lignocellulose feed. Hence, $N = 4$. The second parameter is the fraction of mass and heat streams that are known, $F$. Given the extensive development program carried out with the pilot plant and due to the presence of a dedicated research test plant, the value of $F = 1$. (The range of $F$ is from 0 to 1.)

The third parameter is the type of feed, $S$. If the feed is a crude solid, then $S = 10.8$ months.

The start-up time parameters and the start-up time of Merrow correlation is then:

$$T_{start-up} = 3.3 + 3.7 \times N - 3.2 \times F + S = 25.7 \text{ months}$$

So the actual start-up time of 27 months is within the uncertainty range of the start-up correlation time. It should be noted that the actual start-up time was taken to be the time to reach name-plate capacity, while the start-up time according to the Merrow definition is the time needed to reach steady state capacity.

It is also interesting to see what measures were taken for a successful start-up using the critical success factors presented by Harmsen [9]. Table 13.1 shows these factors for the EMPYRO case.

So, except for the first, all critical success factors for start-up were in place.

**Table 13.1:** Critical success factors commercial-scale start-up for the EMPYRO case.

| Critical success factor | Empyro |
| --- | --- |
| **Development and design** | |
| Process defined as new? | No. BTG-BTL did not consider the process as new. They viewed the larger scale as the challenge. |
| Integrated down-scale pilot plant available? | Yes. BTG has a test-plants of 5 kg/h and 200 kg/h. |
| R&D and EPC knowledge integration? | Yes, together with Zeton and BTG. Furthermore, BTL has an advisory board with senior executives from Shell, Aker Solutions, and AkzoNobel. |
| Scale-up knowledge unit operation available? | Yes, experience to scale up from 200 kg/h to 2 t/h |
| **Start-up preparation** | |
| Potential problem analysis carried out? | Yes, extensive HAZOP studies have been carried out throughout the complete plant. |
| Precaution measures taken? | Yes. Client Friesland Campina can operate furnace with Natural Gas, if bio-oil is not delivered. |
| Complete start-up team? Operators trained for start-up and operation? | Yes, by the operators, Empyro plant management, and BTL people Yes, training for half a year, both theoretical and practical |
| Start-up plan? | Yes, detailed plan was available and every unit has been started up separately. |
| Documentation? | Yes, detailed documentation was available about the complete plant. |

## 13.8  Case evaluation

### 13.8.1  Sustainable development goals of BTG-BTL

The BTG-BTL vision and mission statements [10] state that it wants to contribute to a more sustainable society by providing a renewable energy alternative to fossil fuels. This will be achieved through the delivery and deployment of fast pyrolysis technology for converting green biomass and organic residues into pyrolysis oil in an environmentally, socially, and economically sustainable manner.

The economic need for a renewable oil, not competing with food or with land for food, is increasing steadily. Pyrolysis liquids can be stored, traded, transported, and applied universally just like mineral oil. Pyrolysis thereby allows the decoupling of biofuel production and its actual use. Hence, it is clear that the technology providing company wants to contribute to the sustainable development goals 7 and 13 (see Chapter 3, sustainable development goals).

### 13.8.2 Role of PI concept design and modeling

This process has several Process Intensification features. Each will be evaluated with the PI categories of Chapter 2.

The first PI feature is that the RCR does not need a heat carrying gas, normally used in other pyrolysis processes. This absence of a carrying gas reduces the complexity and size of down-stream processing. Hence, plant-wide spatial PI design is applied.

The second PI feature is that the heat generated in the combustion section is transported back to the reactor by hot sand. Hence, this is a novel way of energy supply to the pyrolysis reactor, reducing the number of process steps as no heat exchanger is needed. Hence, again a PI plant-wide process intensification category design is applied. Also the PI design category, alternative way of providing energy to the pyrolysis reaction step, is applied.

The third PI feature is that the biomass particles are heated up rapidly and uniformly in the high shear RCR, while moving upward along the cone. The biomass particles have a short plug flow residence time of 2 seconds, resulting in a high oil yield of typically 70% on dry biomass intake. Hence, this is a unit operation intensification, using plug flow and high heat transfer phenomena.

The fourth PI feature is that both char and off-gas are combusted in the same reactor. This is PI design using function integration.

The fifth PI feature is that the surplus energy of the combustor is converted to high pressure steam. This steam is partly used to dry the fresh biomass, partly for steam export to other processes or district heating nearby, and partly for electricity generation. Hence, this results in energy integration using a very efficient breakthrough PI process.

The sixth PI feature is that the whole process design is a skid-mounted design consisting of 22 modules. This results in a compact construction that is easily transported to the production site. This ensures that the investment costs are kept low. Dry testing is carried out at the construction site, resulting in a short commissioning time at the production site with little additional operator support.

In summary, four of the five PI design categories are applied in this case.

### 13.8.3 Role of modeling

Reactor and process modeling played a minor role in the process design and scale-up. The main focus was on experimentation and various scale-ups. However, reaction kinetics and heat transfer modeling played a role in the reaction pyrolysis and development of the combustion concept design.

Dynamic process modeling is now underway for operator training.

### 13.8.4 VIB perspectives

The shared values of engineers that are integrated in this case from idea to commercial-scale operation are wanting to be innovative and wanting to have the invention in commercial-scale operation The engineers involved are all from the same Twente region and several of them were involved from the start and are still involved.

The stakeholders involved in the Empyro project are listed below.
1)  BTG Biomass Technology Group BV The Netherlands BTG
2)  AkzoNobel Base Chemicals BV The Netherlands AkzoNobel
3)  Amandus Kahl GmbH Germany AK
4)  Bruins & Kwast Recycling BV The Netherlands BKR
5)  R&R Consult Denmark RRC
6)  Stork Thermeq BV The Netherlands STQ

Their interests can be summed up as having a promising innovation that generates employability for the Twente region. The governing belief of the local society is that local co-operation is important for developing this region. This belief is latently present in the engineers, the stakeholders, and the local government.

### 13.8.5 Multidisciplinary cooperation

Before describing the multidisciplinary cooperation, it is important to first describe the breakthrough innovation aspects of this case. This case is about:
–  A new market (pyrolysis oil for industrial heating) – this market did not exist
–  A new product (stable transportable biopyrolysis oil)
–  A radically new process

Combining these three new elements in a project is, in general, a route to commercial-scale disaster [9].

This innovation case was still very successful, going on to becoming the world leader in biomass pyrolysis technology that produces stable bio-oil (after the successful EMPYRO startup, two new projects are under construction in Finland and Sweden). This success is due to a very special multidisciplinary cooperation (in combination with the other factors mentioned in Sections 18.8.1–18.8.3).

Multidisciplinary cooperation, in this case, is executed as follows. The director of BTG-BTL has a formal education in electrical engineering and has a long experience in marketing and sales of technology in Stork engineering company, Wartsila Diesel, and Holec Switchgears. When he became the director of BTG-BTL, he used his tacit knowledge to lead further development of the technology, get co-operation from the EPC companies, find shareholders for EMPYRO, and market the bio-oil, that is, find a client for this new product. Hence, a key part of the multidisciplinary cooperation happened due to this person. The disciplines involved are: development, development management, engineering knowledge, marketing the technology, defining a new market, and marketing the product.

Transferring the developing engineer to the EPC contractor ensured another multidisciplinary cooperation..

### 13.8.5 Formal education

This section is about the role of formal education in the design, development, and implementation as distinct from learning from experience that was described in the previous sections.

Formal education in chemical engineering and fluidization played a role in the discovery and concept stages, where students were educated by Professor van Swaaij.

Mechanical engineering education, as distinct from chemical engineering, played a major role in making the sand recycle flow reliable from the combustor back to the pyrolysis reactor.

Operator training for startup and normal operation played a very important role. It is facilitated by hands-on training on the integrated pilot plant. Presently, a dynamic process simulator is built to train operators on the new processes to be implemented.

### 13.8.6 Learning points

A very risky innovation project with three risks, a new nonexisting market (bioheating fuels), a new product (stable liquid biofuel from biomass pyrolysis), and a radically new highly intensified process, was in the end very successful.

We, Harmsen and Verkerk, analyzed this success with the theoretical framework of Chapters 2–4. It is likely that this success is due to:

The knowledge base of the van "Swaaij school education" was the same for most of the engineers involved.

There was knowledge transfer over the stages by people moving as knowledge carriers first from the university (discovery stage) to the BTG and later to the BTG-BTL company. This was complemented further by moving a person from the development to the EPC stage. In this way, time-travel back to previous stages happened easily.

Having the development, EPC, and commercial implementation within the same Twente region, a local society with a distinctive culture of getting to know each other and thereby building trust between many different stakeholders ensured that risks were understood and accepted.

## References

[1]   Wagenaar, BM, The rotating cone reactor for rapid thermal solids processing. PhD Thesis Universiteit Twente, Hengelo, Netherlands, 1994.
[2]   Wagenaar BM, Prins W, and Van Swaaij WP. Pyrolysis of biomass in the rotating cone reactor: modelling and experimental justification. Chemical Engineering Science. 1994 Dec 1;49(24), 5109–5126.
[3]   Janse AM, Prins W, and Van Swaaij WP, Development of a Small Integrated Pilot Plant for Flash Pyrolysis of Biomass. In developments in thermochemical biomass conversion, Dordrecht, Springer, 1997, 368–377.
[4]   Janse, AM, A heat integrated rotating cone reactor system -for flash pyrolysis of biomass-. PhD Thesis University Twente, Enschede, 1998.
[5]   Janse AM, de Jonge HG, Prins W, and van Swaaij WP. Combustion kinetics of char obtained by flash pyrolysis of pine wood. Industrial & engineering chemistry research, 1998, Oct 5;37(10), 3909–3918.
[6]   Janse AM, De Jong XA, Prins W, and van Swaaij WP. Heat transfer coefficients in the rotating cone reactor. Powder technology. 1999 Dec 6;106(3), 168–175.
[7]   Westerhout RW, Waanders J, Kuipers JA, and van Swaaij WP. Recycling of polyethene and polypropene in a novel bench-scale rotating cone reactor by high-temperature pyrolysis, Industrial & engineering chemistry research, 1998, Jun 1; 37(6), 2293–2300.
[8]   Venderbosch, 2017, Sourced 13-12-2019 http://www.umb.no/statisk/2/Bio4Fuels%20presen tations/Venderbosch.pdf.
[9]   Harmsen, J, Industrial process scale-up – a practical innovation guide from idea to commercial implementation, 2nd revised edition, Elsevier, Amsterdam, 2019.
[10]  BTG-BTL, 2019, Sourced 13-12-2019 www.btg-btl.com.

# 14 Eastman chemical methyl acetate multifunctional integration

## 14.1 Background information

The Eastman Chemical Corporation has developed and implemented a breakthrough process-intensified methyl acetate process. The core of the commercial-scale process is a multifunction integrated column. The functions integrated are reaction, solvent extraction, vapor stripping, and distillation. The commercial-scale column is over 80 m high, has a diameter of 4 m, and has a production capacity of 200,000 t/year [1]. The process start-up was in 1983 [2]. A few years later, a second column was installed for further capacity increase [1, 3].

The reaction step for methyl acetate is a classic esterification of an acid reacting with an alcohol, to form an ester and water. Liquid acetic acid in the presence of an acidic catalyst (sulfuric acid) reacts with liquid methanol to form methyl acetate and water. This is an equilibrium reaction, which means that total conversion cannot be reached in a single pass reaction. The products are to be removed from the feeds, and the feeds are to be recycled back to the reaction section. That separation is also not simple as the products and feeds, all form azeotropes. This means that several separation steps with recycles are needed to obtain the pure product streams. The conventional process design made by Eastman required two reactors and eight distillation columns [4, 5].

Just prior to start the detailed engineering design, the engineers in Eastman decided to go for a single reactive distillation column design; an idea they had been toying with, for several years. This meant not only a new design, but also going through all development steps, while only a year was available before entering the engineering procurement and construction stage [2].

I investigated how Eastman came to this highly intensified design, how they did the scale-up, and how values of engineers, interests of stakeholders, and beliefs of the local society played a role in this remarkable innovation, by gathering all publications by Eastman engineers. In addition, I interviewed Jeff Siirola [3]. Jeff had obtained his PhD on the subject of process synthesis in 1970 [6], and then joined Eastman Chemical in Tennessee [3]. He continued to work on developing the process synthesis method [7, 8] and applied it to the Eastman methyl acetate process design. He had also created a process innovation method [4, 5]; hence, he was the pivotal person to be interviewed.

The information in the papers and the results of the Siirola interview are placed in appropriate sections of this chapter. For clarity, the stages structure of this book is used. Deviations of that structure from the Eastman innovation pathway are noted.

https://doi.org/10.1515/9783110657357-014

## 14.2 Discovery stage

### 14.2.1 Discovery stage introduction

The route from idea to commercial-scale implementation is, for the Eastman case, not a simple linear step-by-step procedure from discovery to end of development. In reality, it was a tortuous route, also containing a back to the beginning step. Ideas of having a reactive distillation process in a single column were considered for some time. The first attempt failed because the formation of an azeotrope was not recognized [2]. Then, a conventional process design was developed resulting in 10 unit operations. This design is shown in Figure 14.1.

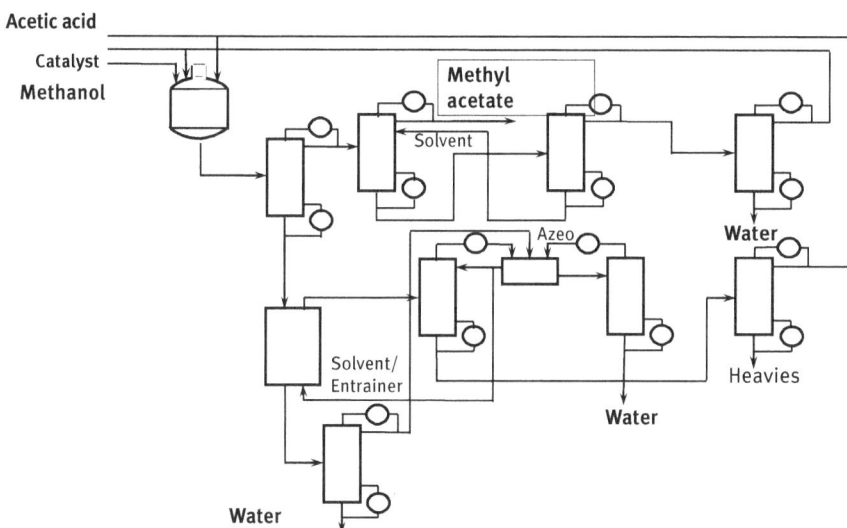

**Figure 14.1:** Conventional process design methyl acetate [4, 5].

Just before the detailed engineering would start, it was decided to go back to the drawing table and make a process-intensified design based on function integration, because the engineers considered the conventional process design to be very complex. This decision was taken despite the fact that there was a time constraint; a delay in the start of the detailed engineering would be expensive and negate any savings of the new design [2].

Jeff Siirola told me how this happened in some detail in the interview of 2019. The leader of a detailed engineering group in Eastman became seriously ill. Jeff, was working in R&D at that time, and was asked to lead that group as interim manager. When he led that group, he sent a few process engineers to the university to be trained in process synthesis, based on function identification and function integration –

a PI design method partly developed by Siirola. When they came back, they immediately started to work on the methyl acetate process design, using the process synthesis method of function identification and integration. They managed to get a process concept design combining all functions in a single column. That concept was then transferred back to the R&D department to be tested and developed [3].

Siirola not only developed a process synthesis design method, but also an innovation method starting from need identification and ending in normal operation of the commercial-scale process. A summary of the latter method is found in [4, 5]. The method has many steps. The first step requires little but essential information. With each step, more information is needed and generated. That method evolved from his experience in process innovation at Eastman. Some of that method and its terminology were already available and used for the methyl acetate case.

## 14.2.2 Discovery stage problem definition

Siirola calls the first step in process innovation, the "targeting step." It involves need identification, [4, 5] in this case, the product market requirement and specification. Agreda describes that the manufacture of high-purity methyl acetate is difficult because of reaction equilibrium limitations and the formation of methyl acetate-methanol and methyl acetate minimum boiling azeotropes [2]. Conventional processes use schemes with multiple reactors in which the large excess of one of the reactants is used to achieve the high conversion of the other reactant. A series of vacuum and atmospheric distillations is needed to change the methyl acetate–water azeotrope. The refined methyl acetate is obtained by separation from the surplus reactants that are recycled back to the reactors. Other process schemes contain an extractive agent to separate methyl acetate from methanol. As remarked before, early attempts to apply reactive distillation failed because the azeotropes were not recognized. Later work found that the formation of methyl acetate–methanol azeotrope could be circumvented by reactive distillation, but concluded that pure methyl acetate cannot be obtained in the primary reactive distillation and that a second column is needed to fractionate methyl acetate from the methyl acetate-methanol azeotrope. Additional problems are posed by the presence of impurities in the feedstock methanol and acetic acid, which may be intermediate boiling compounds that will accumulate in the process, contaminating the product, and therefore require additional columns for removal [2].

In 1980, Eastman first chose a conventional methyl acetate process-design as part of a grand plan for the development of plants from coal as primary feedstock, meaning that it also contained plants for the production of the feedstock methanol and acetic acid. The design consisted of two reactors and eight distillation columns see Figure 14.1. Development work including a pilot plant had also been completed.

However, due to the complexity of the plant, it was deemed desirable to pursue the search for a more economical process, in spite of the fact that detailed engineering and the procurement of materials for the plant was about to begin. Delays in the latter could be quite expensive and negate any savings due to optimization work [2].

### 14.2.3 Design, modeling, and testing

Many process design options obtained from literature and patent reviews on esterification processes, in general, and methyl acetate, in particular, were tested by computer simulations. The computer program necessary for these simulations was developed within Eastman. By evaluating design options with the models, the single reactive countercurrent extractive distillation methyl acetate reactor column was obtained [2].

No experimental testing is reported by Eastman for this discovery stage. The proof-of-principle testing seems to have been done by computer simulation only.

## 14.3 Concept stage

### 14.3.1 Concept stage content of Eastman case

Siirola calls the concept stage the preliminary level (the next stage after the targeting level) [4, 5]. The purpose is to arrive at a tentative, not-too-detailed design solution to the given innovation problem.

### 14.3.2 Concept design

Agreda describes the concept stage in some detail. The reaction is an equilibrium reaction. The partition coefficient of the individual reactants and products are such that, in the vapor phase, the products to reactant ratio is higher than in the liquid phase. As such, in reactive distillation, a higher conversion can be obtained than in a single liquid phase reactor.

The reaction kinetics is modeled as: $r = k_0 \, e^{-E/RT} \, [C_{\text{MeOAc}} \, C_{\text{H2O}} - C_{\text{HOAc}} C_{\text{MeOH}}/K]$ in which $k_0$ is the reaction rate constant, that is a function of the sulfuric acid concentration. The expression is strongly nonlinear and only holds for a certain concentration range. In general, acids in organic media are strongly nonlinear and also strongly depend on water content. Therefore, that expression is not provided here. $K$ is the equilibrium constant for the reaction, which dependents on temperature. Further quantitative information on parameters is not provided by Agreda [2].

Due to the reaction equilibrium, the reaction rate in the liquid is increased by removing one of the products (methyl acetate preferentially) from the other components, by evaporation. Furthermore, if the reactants can flow countercurrently, high concentrations at opposite ends will ensure high conversion of both reactants. Thus, the concept of a series of countercurrent flash reactors can be used to great advantage, in this particular reaction [2].

Separation by distillation has several points of concern. Methyl acetate forms a minimum boiling azeotrope with water, with atmospheric boiling point $T_{boil}$=56.1 °C. It also forms an azeotrope with methanol with boiling point $T_{boil}$=53.9 °C. The boiling temperature of methyl acetate itself is $T_{boi}$= 57 °C, very close to the azeotrope boiling points. Therefore, conventional distillation design requires several vacuum distillation columns to change the boiling points. An alternative is to use extractive agents to avoid the azeotropes. Eastman engineers discovered that acetic acid is such an agent, and hence, can be used to obtain pure methyl acetate [2].

Siirola describes first, in general terms, this early design approach which he calls "means to ends" with tasks process synthesis and also evolutionary modification [5]. It is a combination of heuristic rules and computer simulation to study the feasibility of intermediate solutions. The flow sheets generated are compared, the most promising is selected for further modification, and streams are considered for extraction and stripping. Siirola then describes a hierarchy in this step-by-step approach. The hierarchy is task identification (what is to be done) and task integration, in that order. Equipment design was postponed to a later stage.

Within his evolutionary concept design method, he further describes a hierarchy of first identity (what to be done), then amount (sizing of extend of task), then concentration, phase, temperature and pressure, and finally, form. The logic of this hierarchy is that properties later (lower) in the hierarchy can be more easily manipulated to arrive at the desired end solution.

In his method, tasks are represented as blocks connected by streams. In the last step, tasks are combined in a single reactive distillation column. Siirola makes it clear that the hierarchy of steps is not dogmatically followed and iterations to previous steps have been applied.

Siirola also describes the methyl acetate concept design of the reactive distillation column, in some detail [5]. Figure 14.2 shows the functions (called "tasks" by Siirola). It is to be noted that not only the functions, but also the process streams play an important role. Streams are used for solvent extraction and for vapor stripping. Figure 14.3 shows how the functions are combined into a single reactive extractive distillation column.

Task G is a simple distillation to obtain pure methyl acetate vapor. Acetic acid flows as liquid to Task F. In this section, the liquid acetic acid extracts water from the vapor flow resulting in a vapor stream upwards of the main product methyl acetate, with some acetic acid. In Task E, a reaction which converts methanol deeply (so it is a methanol removal step) takes place. Then, in Task A the main reaction of

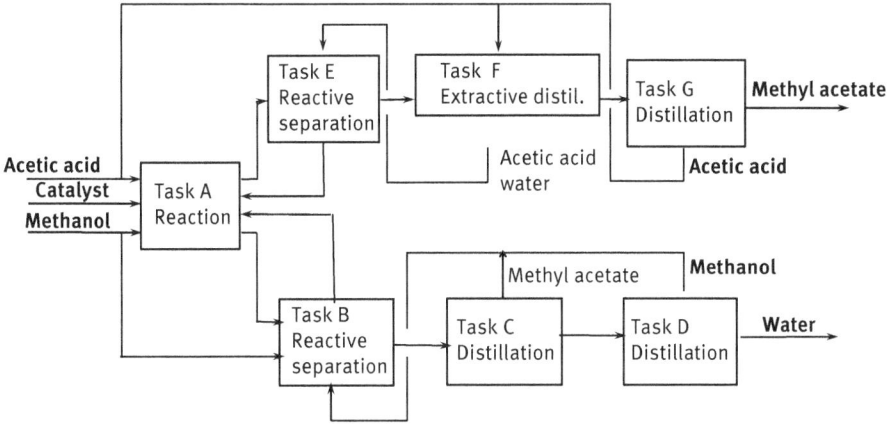

**Figure 14.2:** Methyl acetate task integration and use of streams [4, 5].

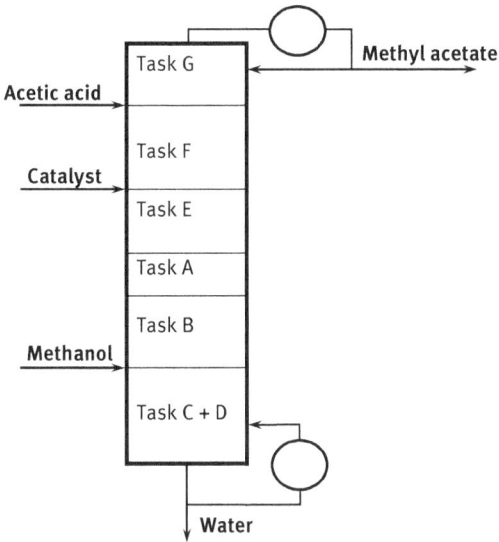

**Figure 14.3:** Function-integrated column methyl acetate process [4, 5].

methanol with acetic acid takes place. Below that, in Task B, acetic acid is removed from the liquid phase by distillation. In Task C, the methyl acetate–water azeotropic mixture is removed as vapor. Finally, in the bottom section, Task C+D, methanol is distilled from the aqueous phase. Water leaves the process at the bottom (together with the sulfuric acid catalyst, and by-products) [2, 5].

### 14.3.3 Concept modeling testing and evaluation

Furthermore, distillation modeling and laboratory-scale testing showed that using acetic acid as extractive agent helped overcome the formation of azeotrope and create pure methyl acetate. The countercurrent extraction could be obtained in the reactive distillation column.

Two options were available for catalysts: heterogeneous Amberlite 200 acid ion-exchange catalyst and homogeneous acidic catalyst. The Amberlite requires a complex mechanical internals design. This alternative was therefore rejected. Several homogeneous catalysts were tested at laboratory reactor- scale. The test showed that sulfuric acid was most suitable with a sufficiently low reaction time when used in reactive distillation. The test also showed that the liquid hold-up would be a critical design factor.

A new economic evaluation with all inputs from the concept design and laboratory test results showed favorable results and larger scale testing was recommended [2].

No formal concept stage-gate is reported by Agreda [2]. Siirola mentioned that some form of stage-gate was applied, but because of close proximity of knowledgeable management, this stage-gate was quickly executed [3].

## 14.4 Feasibility and development stage

### 14.4.1 Feasibility and development approach

A formal feasibility stage away from the development stage is not described by Agreda [2]. Therefore, feasibility and development are treated here together in this section.

### 14.4.2 Experimental bench-scale testing

The process development started with a bench-scale reactive distillation with a capacity of 2.3kg/h of acetic acid in a glass column with a diameter of 10.2cm (4 inch) and a height of 9.14m. It also contained a high liquid hold-up section for the main reaction. Feed representative for the commercial- scale process was then already used [2].

Operating this column revealed that the reflux ratio was a very important factor in the process. It could however not be determined with available literature knowledge. It also revealed that process control resulting in both high purity methyl acetate as top product and water as by-product in the bottom was hard to achieve, in combination with a stoichiometric feed of acetic acid and methanol. It was decided that a larger pilot plant would be needed to study the control issue in more detail [2].

Also, corrosion testing was started in the bench scale unit and the steel type (Carpenter 20 Cb-3) was chosen as construction material for the commercial-scale column [2].

### 14.4.3 Pilot plant design and test results

Agreda mentions that the main purpose of the pilot plant is to confirm the feasibility and controllability of the process over a long time, and furthermore, to provide design data needed for full-scale implementation. Agreda mentions, in particular, the pilot plant purpose to demonstrate the reactive distillation column capability to produce methyl acetate with excellent quality over a significant period of time, at high yields, and in such a manner that process control and recovery from upsets could be performed easily, without the use of complex control equipment (the reader is reminded that the pilot plant was built in between 1980 and 1983) [2].

Agreda discloses that the pilot plant needed a much larger diameter (compared to the bench-scale unit) to ensure adiabatic operation of the column, which is essential for a proper function of the pilot plant. Adiabatic measures for reactive distillation pilot plants are very important for obtaining reliable information for scale-up. If heat is leaving via the column wall, then not all reaction heat is going to vaporization. The vapor flow and liquid reflux flow of the pilot plant are then different from the commercial-scale plant for unknown reasons. Hence, basing the commercial-scale plant design on nonadiabatic pilot plant information will result in a disaster.

The pilot plant had a capacity of 45kg/h (360t/year) with a diameter of 20cm (8 inch) in the stripping and reaction sections, and 15cm (6 inch) in the extraction and rectification section. The column height was 30.5m (100ft). This reactor had high liquid hold-up, reverse flow bubble-cap trays in the reaction and stripping sections, and Pall ring packing in the extraction and rectifying sections. Special precautions were taken to ensure adiabatic operation (of the column excluding the bottom boiler and top cooler) for accurate reflux measurements.

The pilot plant also contained an intermediate boiling impurity system of a stripping column and concentration column. This system allowed the recycle of methyl acetate and unreacted acetic acid from the side draw stream, back to the reactor column.

Detailed corrosion studies were made of the best candidates for construction materials for all equipment that would be needed for the commercial-scale plant.

A detailed process control scheme was also made for the whole pilot plant, as well as sample points and on-line analyzers to determine tray and stream compositions to obtain information for the commercial- scale plant.

The pilot plant results and a new economic analysis confirmed the feasibility and the economic advantages of the process. The reactive distillation column process was then chosen for full-scale implementation [2].

### 14.4.4 Concurrent modeling and design in the development stage

The entire development, from concept inception to the start of the detailed engineering design for the commercial-scale plant, was completed in less than a year. This very short time span was obtained by having several tasks carried out concurrently. Before experimental tests were completed at a certain scale, design of the next scale was started, and also commercial-scale designs and their modeling and simulation were carried out concurrently. Also, the detailed engineering design started before the end of the pilot plant testing [2].

Agreda also reports that at the time the concept proposal was made, Eastman had not yet developed a reactive distillation simulator. Given the urgency, stage-to-stage hand calculations were performed using a flash-with-reactor program. Distillation sections were simulated with an existing distillation program [2].

Concurrently with the pilot plant work, two reactive distillation computer programs were developed – the first being a dynamic program and the second, a steady-state simulator [2]. Hence, concept design, computer program development, modeling, bench-scale testing, and pilot plant design were done concurrently and within one year.

This enormously speedy concurrent innovation could be so fast, only because all people involved had the same goal in mind and worked in close proximity from each other [3]. Siirola mentioned in the interview that all R&D and detailed design engineers were in the same building and only yards from each other. They knew each other very well; hence communication was easy and fast. The decision to pass the stage-gate to the engineering procurement and construction (EPC) stage was easily taken, as the top manager was a chemical engineer, who understood how the R&D engineers had tested the design experimentally; he also understood why a scale-up from the pilot plant by a factor of 500 could be done reliably [3].

## 14.5 Engineering procurement and construction

Detailed engineering work started when the pilot plant development neared completion. Design heat and material balances were prepared, column and tray designs were finalized, and process and instrumentation diagrams (P&ID) were prepared. The commercial-scale design consisted of the reactive distillation column and three distillation columns, that is, one column more than the pilot plant. A methanol recovery column was added to process the underflow of the reactive distillation column for two reasons:
1) To lower the reaction base temperature, thus reduction corrosion rates
2) To provide minimum methanol losses during process upsets.

The detailed engineering also helped make decisions on the reactive column design, which, in hindsight, appeared to have a negative impact on the operation; see the next section on implementation. The detailed design engineers, being aware of the additional cost of a much higher tower, had taken this decision. Agreda, in hindsight, was sorry for this decision as it left no operation freedom for the reflux ratio at design capacity [2].

The plot plan and construction drawings were completed, and procurement and construction began before the pilot plant test was completed. No further information is available on procurement and construction.

The whole EPC stage including precommissioning took less two years.

## 14.6 Implementation start-up and normal operation

During the construction phase operating procedures were written for start-up, normal operation, normal shut-down emergency shut-down, trouble shooting, rate changes, and so on [2].

Upon completion, the plant inspection and water check-up took place. Many problems were encountered and corrected. Problems included improper equipment installation, construction debris blocking pipes, incorrect sizing, and incorrect calibration of equipment and instrumentation [2].

After this check-out, the plant started up in May 1983. The plant was the first to start-up in the whole complex of coal to chemicals site [2]. The most challenging debugging and troubleshooting took place when the production rate was raised to design capacity of 22,700kg/h (200,000t/year). The column height, reported to me by Siirola for my publication [1] was over 80m and the column diameter was 4m.

Processing limitations were eliminated as they were identified. In some instances, the limitations came from outside: by insufficient cooling tower water from the site facilities, which also had start-up problems. Other limitations were due to improperly sized equipment, such as pumps and condensers. However, the most difficult problems to correct were related to the scale-up and design of the reactive distillation column [2].

The pilot plant results showed that the reactive distillation column could be satisfactorily operated at reflux ratios ranging from 1.5 to 2.0 with an optimum ratio of 1.7. For economic reasons, the commercial scale was designed with a reflux ratio of 1.5. This reduced the flexibility at design capacity operation [2] to zero. Furthermore, each section of the reactive distillation column was designed for the expected steady-stage vapor and liquid loading at the minimum factor for flooding [2].

Finally, due to the high liquid hold-up requirements for the reaction zone trays, the design calculations had to be performed by extrapolating correlations for standard bubble-cap trays, acknowledging the risks of such extrapolations from 5 inches to 12.7 inches (0.32m). Moreover, the design of the down comer sumps was such that it had only a vertical entrance opening of 9inches (0.23m). The liquid on the tray is much more aerated than in normal distillation, by which highly aerated (gas containing) liquid enters the bottom comer, where also more vapor is generated. This caused excessive pressure drop, and flooding occurred. Actual operation had shown moreover that reflux ratios of 1.65 to 1.85, depending on product requirements, were needed [2].

In the end, all problems were corrected by de-bottlenecking the column sections during a normal shutdown and by increasing the column's operation pressure from 3 to 10 psig [2]. This latter piece of information is uncertain as Agreda also states that the operation pressure increased from 122 to 170kPa [2]. Which of the two statements are correct is unclear. My guess is that the latter pressure increase by a factor 1.4 is more likely than the increase by a factor 3.3.

Agreda reports that process improvement and optimization have allowed process operation at up to 125% of design capacity. He does not report how long it took to achieve this increase. Operation is extremely easy as long as it is kept within hydraulic capability limits. In normal commercial-scale operation, the methanol recovery plant and the intermediate boiler distillation columns are often not in operation. Typically, at that time, normal shutdowns were every four years, so the problems reported were probably solved within 4 years. No formal information on start-up time is provided [2].

The energy required and investment cost were 80% less when compared to the conventional design [9]. Siirola reported that, in 1990, a second plant has been started up. The reactive distillation column has modifications on sieve trays in the reaction section [1].

## 14.7 Evaluations: Eastman case

### 14.7.1 Process intensification design, modeling, and validation evaluation

#### 14.7.1.1 PI design evaluation

The Eastman design case falls in more than one process-intensification domain. First, it falls into the spatial plant-wide category, as the whole plant design was reconsidered and not just individual unit operations. In the end, nearly the whole plant from input to output could be designed with two columns rather than 10 separate unit operations.

Second, it falls in the PI domain function synergy as many functions were integrated in a single column. Siirola's task definition and task integration description are the same as function definition and function integration. His design method began from his university research and was improved by extracting knowledge from Eastman case experiences.

This type of general knowledge generation by experienced designers is well described by Schon in his book *The Reflective Practitioner – How Professionals Think in Action* [4]. Siirola's design method is incorporated in Chapter 2, in methods for process intensification concept design with function integration.

### 14.7.1.2 Modeling and validation evaluation

Agreda and Siirola reports reveal that modeling, simulation, and validation played important roles in defining and optimizing the complex reactive distillation solution, combining 6 functions in a single column, with stoichiometric feed of reactants. In fact, finding the final solution could only be done by design thinking, modeling, experimental validation, and repeating this cycle several times.

## 14.7.2 Values, interests, and beliefs evaluation

### 14.7.2.1 Values of engineers evaluation

Eastman went back from the development stage to the discovery stage to explore and find a superior process idea, because the engineers did not like the complex conventional process design. This is a nice example of the power of the value of engineers' perspective. They took this remarkable initiative to find a less complex novel process design, despite the time pressure. The hurdle of needing additional time was overcome, to some extent, by having a concurrent design, modeling, and validation effort, by having concurrency in the discovery, concept, feasibility, development, and front-end engineering design, by explaining to management that a superior solution with respect to cost and reliability had been obtained, and by excellent cooperation between all people involved.

It is also remarkable that Agreda and Siirola reported enormous amounts of detail on the whole innovation of this process. Companies rarely report their innovation in detail. It is clear that Agreda and Siirola felt a strong need to let the engineering community know how they had achieved this breakthrough process innovation with a factor 5 lower investment cost and energy requirements. By their publications, they showed that they value sharing developed knowledge for the benefit of other engineers and the society.

In the interview, Siirola mentioned three values of engineers that played a role in Eastman, and also mentioned that besides fulfilling the constraints on safety, health, environment, and social, the important values were:
1) Having new innovative processes
2) Having reliable production and product
3) Having an economical process

### 14.7.2.2 Interests of stakeholders evaluation

The interest of the stakeholders is not mentioned in the Eastman publications, but Siirola mentioned that Eastman Chemical was the main employer of the town in Tennessee and had not fired any employee (till 1999) since its start in 1920. Everybody in the company knew the importance of the company for employment in the local society. Having new and reliable competitive processes resulted from that responsibility. The interest of clients for the methyl acetate product was taken care by producing sufficient product and the required quality all the time.

The publications mention that all design and development were also focused on a process that produced high purity methyl acetate for the clients, all the time.

Also, the interest of the process operators was probably in mind when striving for a simple and easy-to-control process.

How the top management and shareholders or stakeholders were involved and how their interests were considered is not revealed in the publications.

### 14.7.2.3 Basic beliefs of society evaluation

The basic beliefs perspective of society is obtained from the Siirola interview in 2019 [3]. He stated that the Eastman Chemical complex, Kingsport, Tennessee, of which the methyl acetate process was an important part, had a large societal aspect, providing 13,500 jobs. It was the largest employer of the city, and so provided a large income to the local city community. Tennessee is a relatively poor state. It is therefore even more understandable that the community, in turn, highly valued Eastman Chemical Company. The Eastman engineers were well aware of their importance to the Kingsport community. This may have been an additional driver to have a very competitive process, such that the Eastman Company would maintain its market position in bulk chemicals production, and thereby maintain its contribution to the local community.

## 14.8 Multidisciplinary assessment evaluation

The paper by Agreda showed many concurrent activities of design, modeling, simulation, and experimental testing, resulting in a front-end engineering design within a year from start. There was no waiting for economic and feasibility stage-gate evaluations, and delays were pre-empted by already starting to work on the next stage [2].

In the 2019 interview, Siirola highlighted that R&D interacted with management of Marketing & Sales and Operation easily because they knew each other, worked in the same building, and production was at the same site. R&D engineers involved had a high reputation in management because of other innovation successes. Eastman had a formal stage-gate procedure. As information was freely shared, this procedure worked smoothly and swiftly [3].

It is clear from this information that multidisciplinary cooperation in Eastman was ingrained in the organization. More than any other critical success factor, this explains why this breakthrough innovation could happen so quickly and successfully.

## 14.9 Learning points for each method

The learning points from the Eastman case for each method are that:
1) Plant-wide spatial design followed by function identification and integration is a very good sequence for obtaining a break-through design.
2) Design, modeling, and experimental validation are essential for a complex function integration design.
3) Values of engineers are an enormous driving force for a breakthrough innovation.
4) Knowing the interest of the local community probably helped in persevering with the innovation.
5) Basic beliefs of the local community cannot be distinguished from the other two perspectives.
6) Multidisciplinary cooperation between R&D, M&S, and OPS creates trust and results in rapid decisions in favor of a breakthrough design.

## References

[1]   Harmsen, GJ, Reactive distillation: The front-runner of industrial process intensification: a full review of commercial applications, research, scale-up, design and operation, Chemical Engineering and Processing: Process Intensification. 2007 Sep 1;46(9), 774–780.
[2]   Agreda, VH, et.al., High-purity methyl acetate via reactive distillation, CEP, 1990, 40–46.
[3]   Siirola, JJ, Interview by Jan Harmsen at AICHE Fall meeting, Orlando, 12th of November 2019.
[4]   Siirola, JJ, Industrial applications of chemical process synthesis, In Advances in chemical engineering, Academic Press., 1996, 23, 1–62.
[5]   Siirola, JJ, An industrial perspective on process synthesis, AIChE Symp. Series 91(304), 1995, No 304, 222–233.
[6]   Siirola, JJ, The computer-aided synthesis of chemical process designs. Ph.D. dissertation, University of Wisconsin, Madison, 1970.
[7]   Siirola, JJ, Powers, GJ, and Rudd, DF, Synthesis of system designs: III. Toward a process concept generator, AIChE Journal, 1971 May 1;17(3), 677–682.
[8]   Rudd, DF, Powers, GJ, and Siirola, JJ. Process synthesis, Prentice-Hall, 1973.
[9]   Siirola, JJ, Synthesis of equipment with integrated functionality, syllabus first Dutch process intensification symposium, Rotterdam, Netherlands 7th of May 1998.

# 15 Shell OMEGA only monoethylene glycol advanced process

## 15.1 Background information

The conventional Shell glycol process consists of an ethylene oxide (EO) section and a glycol (G) section. In the EO section, ethylene is partially oxidized to EO. The by-products, approximately 10%, are carbon dioxide and water. In the glycol section, EO reacts with water to form monoethylene glycol (MEG). The by-products are diethylene glycol (DEG), triethylene glycol (TEG), and heavy ends (HE), which together constitute 10% of the total reaction. This amount of by-products results from the thermal reaction system in which EO reacts with water to form MEG. EO, however, also reacts with MEG to form DEG and also with the DEG to form TEG. This by-product formation could only be limited to 10% by supplying a large surplus of water over EO at the ratio of 22 mol/mol such that the first reaction to MEG is faster than the other reactions. This large surplus of water has to be removed by evaporation from the product stream. Also, DEG and TEG have to be separated from MEG and from each other. This separation is carried out by distillation. As glycols are high-temperature boilers, high-pressure steam is needed for heating the bottom section of the distillation columns. This requires large amounts of energy while a large wastewater stream is also created [1].

This conventional process design is licensed by Shell to ethylene glycol manufacturers over 50 times and Shell is the market leader in licensing the ethylene glycol processes.

The new OMEGA process concerns the glycol section. The EO section remains the same. OMEGA is a two-step reaction process. In the first step, carbon dioxide (from the EO section) reacts with EO to form ethylene carbonate (EC). According to stoichiometry:

$$C_2H_4O + CO_2 \rightarrow C_3H_4O_3$$

This reaction is exothermic with a reaction enthalpy of $-24$ kcal/mol.

In the second step, EC reacts with water to form MEG and carbon dioxide. According to stoichiometry:

$$C_3H_4O_3 + H_2 \rightarrow HOC_2H_4OH + CO_2$$

This reaction is endothermic with a reaction enthalpy of 2 kcal/mol.

The reactions are homogeneously catalyzed by a set of catalysts. The by-product formation is less than 0.5%. The EC conversion rate is >99.9999% and hence EC is absent in the reactor outlet stream. A small quantity of water is needed for the reaction. The OMEGA process, therefore, requires a small separation section to remove carbon dioxide and trace amounts of water. The separation section is also

https://doi.org/10.1515/9783110657357-015

needed for recovering MEG from trace amounts of HE and for recovering the homo-geneous catalyst. Figure 15.1 shows the conventional and the new process.

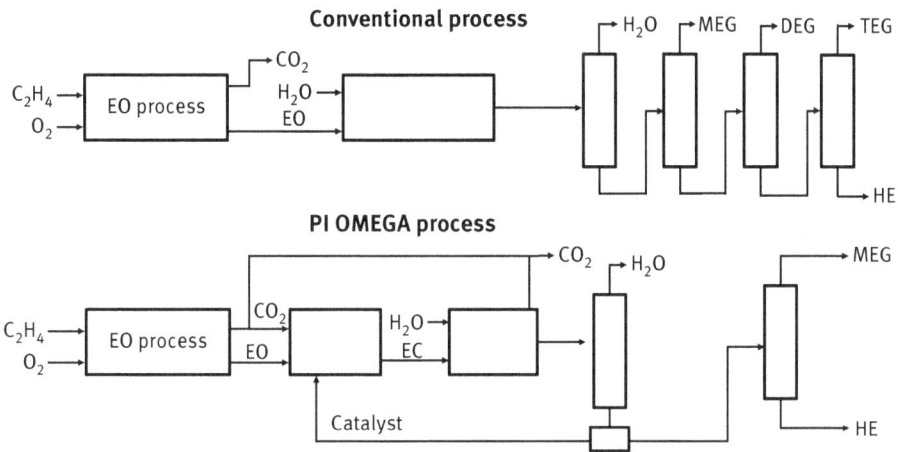

**Figure 15.1:** Conventional and Process Intensified (PI) OMEGA ethylene glycol processes.

The OMEGA process needs 10% less capital investment, 20% less energy, and produces 30% less wastewater than the conventional glycol process [1, 2]. So, the OMEGA process is a good example of the benefits of plant-wide process intensification.

The OMEGA process has been developed by Mitsubishi Chemical Corporation's (MCC) research and development. At the end of its development, the process was licensed to Shell. The author of this chapter, Jan Harmsen, worked in Shell Technology Center Amsterdam at that time, as the focal point at the Enhanced Unit Operations, the Shell term for Process Intensification. Several process experts, including Harmsen, were asked to evaluate the OMEGA process for its suitability for commercial-scale implementation at Shell customers. The insights gained by this evaluation are presented in this chapter.

## 15.2 Discovery stage

MCC started the research in 1996 [2]. It released a patent on the OMEGA process in 1998 stating that: "The present invention relates to a process for producing ethylene glycol from ethylene oxide wherein ethylene oxide in a gas resulting from oxidation of ethylene is absorbed in a specific absorbing solution, is allowed to react with carbon dioxide, converted into ethylene carbonate, and then subjected to hydrolysis to produce ethylene glycol. According to the present invention, a large energy consuming step such as stripping of ethylene oxide and separation of excess amounts of water during the ethylene glycol production becomes unnecessary and the process can be greatly simplified by combining the ethylene oxide absorption step and the carbonation step" [3].

This patent text shows that from the beginning, the whole new glycol process and its advantages were in view. More information on the homogeneous catalyst applied is presented by the inventor in 2010 [4].

## 15.3 Concept stage

In the concept stage, Mitsubishi designed and operated a bench scale process. The process was developed by Mitsubishi from this bench-scale. Kawabe discloses some details of the research on the homogeneous phosphonium catalyst and its reaction mechanism. He also shows the whole process concept in a block flow diagram with the reaction sections, separation sections, and the recycle of the phosphonium catalyst [4].

Kawabe also states that the intermediate component, EC is an important material for lithium batteries [4]. Mitsubishi Chemical now sells high purity EC for applications in Lithium battery cells [5].

Based on this information, it is likely that the business case for the MCC, in the concept stage, rested on the production of this intermediate EC for the lithium battery market and that a lower cost MEG process was an additional advantage.

## 15.4 Feasibility stage

No information is publicly available about what Mitsubishi took into account in the feasibility stage. They decided to design, construct, and operate a pilot plant of 1,000 t/year, followed by the design, construction, and operation of a demo plant of 15,000 t/year [2].

This involves, to the authors estimate, an investment of far more than 10 M€. It is, therefore, likely that this investment was evaluated in the light of future earnings. It may also be the case that the demonstration plant also produced EC which was sold at a profit to customers for the manufacturing of Lithium battery cells for electric cars.

## 15.5 Development stage

### 15.5.1 MCC development

Mitsubishi developed the process further by a pilot plant (1,000 t/year). This was followed by demo plant (15,000 t/year) tests till 2001. The product was tested by a client for producing PET and showed the same behavior as ethylene glycol from the conventional process [2].

## 15.5.2 Critical review process design and scale-up by Shell experts

Shell acquired the exclusive license of the process in 2002. Mitsubishi had developed the process for 6 years [2], providing pilot plant results, demonstration scale results, and a design package.

Shell experts critically reviewed the design and scale-up results of Mitsubishi. The following flaws were identified.

First flaw: The commercial-scale reactor is a crossflow bubble reactor with vertical baffles to reduce the backmixing of liquid such that some staged flow is obtained. The demonstration reactor contained, however, several completely separated reaction chambers connected by pipes for liquid flow. The Shell reaction engineers considered that the commercial-scale reactor design could show some shortcutting flow so that the required very deep conversion would not be obtained. While the demonstration plant has completely separated chambers connected by pipes with turbulent flow hydrodynamics, the latter would prevent the short-cut flows by intimate mixing. The Shell experts asked Mitsubishi to calculate the residence time distribution of the commercial-scale reactor with their CFD model using the particle tracking mode. The model indeed showed some short cutting flow. The commercial-scale design baffles were then modified with horizontal liquid flow sections so that short cutting was avoided and a patent was issued [6]. Figure 15.2 shows the new baffle design. The CFD model now showed that shortcutting was prevented.

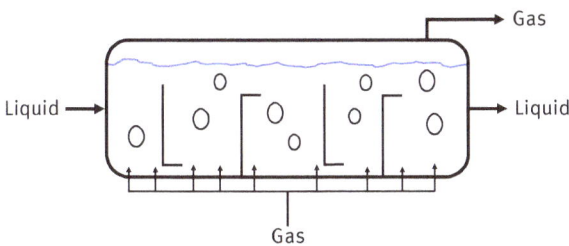

**Figure 15.2:** Reactor with new baffle design.

Second flaw: The demonstration plant did not need any homogeneous catalyst addition to maintain the desired conversion in the very long demonstration run. Mitsubishi concluded that the catalyst had shown no deterioration. Shell experts, however, were not convinced as a large liquid hold-up vessel was included in the catalyst recycle stream of the demonstration scale. So, a large surplus of catalyst was present in the demonstration scale. The commercial scale, however contained, a much smaller buffer vessel relative to the production capacity. Upon a deeper analysis of the recycle stream by MCC, it appeared that significant catalyst deterioration had taken place in the demonstration scale, requiring catalyst addition during the commercial-scale operation.

These findings stress the point that a pilot plant and demonstration plant have to be down-scaled versions of the commercial-scale plant for all aspects critical to scale-up [7]. Thus, having an outsider to review the scale-up helps minimize errors. [8]

## 15.6 The detailed design (engineering procurement and construction) stage

The first commercial-scale process for 400 kt/a was designed and constructed by South Korea's Samsung Engineering, in close cooperation with Shell. It started in January 2006 with a kick-off meeting. The basic design package was ready in July 2006. Mechanical completion was achieved by March 2008 [2].

The second and third commercial-scale processes were designed in parallel to the first process. Table 15.1 shows an overview of these designed and implemented processes.

**Table 15.1:** Commercial-scale implementations OMEGA process [2, 9–11].

| Year | Location | Capacity kt/a |
|---|---|---|
| 2008 May | South Korea | 400 |
| 2009 April | Saudi Arabia | 600 |
| 2009 November | Singapore | 750 |

## 15.7 Start-up stage

Start-up of the first commercial-scale plant started on May 21, 2008 and the guaranteed test run was completed on June 1, 2008 [2]. This means that the start-up time was 10 days (0.3 months).

This start-up time of 0.3 months compares favorably with the start-up time predicted from the Merrow start-up correlation [7] of 3.8 months. The prediction has been obtained by inserting for New process units N, N = 1 for the new reaction section and inserting for the number of mass and heat flows F = 1 as all flows are known from the process model and validated by the demonstration plant. It is likely that for the start-up also all preparation measures have been taken as published by Harmsen [7], as the Merrow start-up time can then be reduced by a factor of 10.

The start-up of the second plant also went smoothly. Arthur Rots, Design Group Leader for Ethylene Oxide/MEG in Shell Global Solutions International BV, managed the project from its inception, working closely with the Petro Rabigh team for more than four years: "We are delighted to see the second OMEGA plant in commercial operation. Petro Rabigh has done an excellent job in preparing for the start-up. The start-up's success reconfirms the strength of the Shell OMEGA design." [2]. From this it is clear that the start-up preparation and execution was carried out by the operation company in close co-operation with the license provider, Shell.

## 15.8 Evaluation OMEGA innovation

### 15.8.1 PI concepts of OMEGA design evaluation

The OMEGA process is a good example of a plant-wide spatial domain Process Intensification design. By considering the whole process, the main inefficiencies of the conventional process, i.e. large by-product formation and consequently a large separation section, have been addressed by a new catalyzed reaction system, producing only the mono glycol product. This reaction section is larger than the conventional reaction section but the intensification is achieved in the separation section due to the lower by-product formation in the reaction section.

The new process design has lower feedstock cost, lower capital cost, lower energy requirements. and lower wastewater production. Hence, it is a prime example of plant-wide wide process intensification with economic and ecological benefits.

### 15.8.2 Sustainable development goals evaluation

At the start of OMEGA process research and development in 1996, the sustainable development goals were already known for 9 years from the Brundtland report published in 1987. In 2002, The UN assembly adopted the Triple P (People, Planet and Profit) definition. Shell has already explicitly committed itself to sustainable development in 1997 and had embraced the Brundtland definition [12]. However, in their reporting on the OMEGA process implementations in 2008, there is no reference to their commitment to the Sustainable Development. Only the benefits to the environment by reduction in energy requirement and wastewater are explicitly mentioned. So, only the planet part of the sustainable development is covered. However, even this is not explicitly stated. It appears that the adoption of contributing to Sustainable Development has been decided on the top level in 1997, but it still had not come to the engineer's level even nine years later. This may be an indication of barriers inside the company between the top management and the level of engineers, in this aspect.

### 15.8.3 Values, interests and beliefs evaluation

#### 15.8.3.1 Values of the engineers evaluation
The MCC and Shell engineers appeared to share the same value of wanting to have a very efficient and reliable commercial-scale process. Proposals by Shell engineers to analyze the potential problems of having EC in the product due to a short-cut flow in the MEG reactor and to remedy it with a baffle modification were immediately accepted by the MCC engineers and a follow-up communication was quickly provided.

This OMEGA process appeals to the values of engineers also because of its highly efficient use of feedstock. In a publication about the OMEGA implementation, Arthur Rots writes as follows: "The holy grail for a process engineer could be the development of a technology that converts all the raw materials to the desired end product with the minimum theoretical energy consumption, no emissions, and the lowest capital cost" [2].

#### 15.8.3.2 Interests of stakeholders evaluation
The interests of stakeholders are first revealed by Shell marketing people listening to their customers and secondly by involving the technology provider MCC. Shell is the market leader in selling ethylene glycol plants and thereby has intimate contact with (potential) customers [10]. A potential customer of Shell said that they wanted a MEG process only from them rather than from MCC, as Shell had a large experience in licensing glycol processes and MCC had not. Due to this interest, Shell decided to contact MCC to obtain a process license and it quickly became apparent that both companies would benefit from this co-operation. Also, the interest of the engineering procurement and construction (EPC) contractor is incorporated by the close co-operation already in the front-end engineering design (base package) as revealed by Arthur Rots of Shell in Section 15.7.

#### 15.8.3.3 Beliefs of society evaluation
The relevance of beliefs of society is analyzed by looking at the countries in which the OMEGA process is first implemented. The new OMEGA process was rapidly implemented in Singapore, South Korea, and Saudi Arabia. All three countries have a strong belief in advancing their societies using the latest technology that requires less energy and has the lowest environmental impact. The OMEGA process is a breakthrough technology. In 2008, OMEGA won an Institute of Chemical Engineers

engineering excellence prize. OMEGA was a finalist for "Commercial Technology of the Year" in the Platts Global Energy Awards 2009 [11].

### 15.8.4 Multidisciplinary cooperation evaluation

The author knows that the marketing department of the Shell glycol process business has a very good connection with the process engineering and the research department. Also, the relations with the glycol process operations are well established and enhanced by personnel exchange between operations and marketing.

This co-operation is due to the continuous EO reactor catalyst sales, catalyst improvements by research, and the consequences for improved process design. Due to the existing connections between these departments communications are smooth. When the customer requested the MCC OMEGA process from Shell, negotiations were quickly started and exchange of technical information between MCC and Shell engineers was immediately established.

## 15.9 Learning points for each method

### 15.9.1 Learning point use of sustainable development goals

Despite the Shell's top management adopting the SDGs nine years earlier, it is still not integrated with the levels of the lower management and engineers. Additional effort would have been needed to achieve this integration.

### 15.9.2 Learning point PI design use

This OMEGA process could only be invented by considering plant-wide process intensification rather than intensifying individual process elements. So, the learning point is, start with plant-wide process intensification first.

### 15.9.3 Learning point values of engineers

When discussing the process design and development, the values of engineers of MCC in Japan, and Shell in the Netherlands appeared to be the same across cultural barriers of wanting to have an efficient and reliable process. Meeting and discussing appeared to be a good way of recognizing this shared value, which resulted in a seamless co-operation. The learning point is then that meetings and discussions between engineers are essential to become aware of shared values and building trust.

### 15.9.4 Learning point interests of stakeholders

The rapid commercial-scale implementation of three OMEGA plants by Shell could only occur due to the knowledge of the customers' needs and in combination with good co-operation with outside stakeholder, MCC, and the inside stakeholders of Shell and MCC, the R&D departments. Silo thinking, or the not invented here syndrome, would have hindered this rapid implementation.

### 15.9.5 Learning point beliefs of society

The first implementation of this breakthrough process was in Southeast Asia. Several breakthrough processes were first implemented in Southeast Asia such as the biomass pyrolysis process of Chapter 14, this OMEGA process, and also the Shell Gas-to-Liquids process. Hence, it appears that the society culture of this part of world is about being receptive to new process technologies. A firm conclusion on this subject, however, cannot be drawn. Such a conclusion would need a wider and deeper study. A starting point can be a study by Albuloushi on the influence of the cultural aspect of uncertainty avoidance on supply chain coordination [13].

### 15.9.6 Learning point multidisciplinary cooperation

Multidisciplinary cooperation between marketing, R&D, engineering, and operations appeared to be essential for buying the breakthrough technology and for rapid implementation at three customers in a very short time.

## Exercises

### Exercise 1: Energy integration
What is the adiabatic temperature rise of the EC reaction?
Q1: Is cooling needed?
Q2: Can the reaction heat be used elsewhere in the process?

### Exercise 2: Sustainable development goals
The EO reaction produces carbon dioxide in high purity in large amounts.
Q1: What are options of using this carbon dioxide as a feed to other applications?

### Exercise 3: Process intensification
Q1: Which functions can you identify in the OMEGA process?

# References

[1] Mitsubishi Chemical, MEG (Catalytic MEG Process) Technology, 2019. Sourced 7 Aug 2019 http://www.mcc-license.com/technologies/pdf/Introduction_MCC_OMEGA_Process.pdf.

[2] Shell,Shell's OMEGA MEG process kicks off in South Korea, ICIS report 2008, Sourced 9 Aug. 2019 https://www.icis.com/explore/resources/news/2008/08/18/9148176/shell-s-omega-meg-process-kicks-off-in-south-korea/

[3] Kawabe, K, Murata, K, and Furuya, T, inventors; Mitsubishi chemical corp, assignee, Ethylene glycol process. United States patent US 5, 763, 691. 1998 Jun 9.

[4] Kawabe, K, Development of highly selective process for mono-ethylene glycol production from ethylene oxide via ethylene carbonate using phosphonium salt catalyst, Catalysis surveys from Asia. 2010 Sep 1;14(3–4), 111–115.

[5] Mitsubishi Chemical, High Purity Ethylene Carbonate, 2019, sourced 6 Aug 2019, https://www.m-chemical.co.jp/en/products/departments/mcc/c2/product/1200986_7910.html.

[6] Harmsen, GJ, and Rots, AWT, EC C07C29/12, 2009, Process for the preparation of alkylene glycol.

[7] Harmsen, J, Industrial process scale-up, 2nd revised edition – a practical innovation guide from idea to commercial implementation, Elsevier, Amsterdam, 2019.

[8] Verkerk MJ, 2019, Visscher F, Industrial Practices, Sustainable Development and Circular Economy: Mitigation of Reductionism and Silo Mentality in the Industry. In The Normative Nature of Social Practices and Ethics in Professional Environments, 2019 IGI Global, 56–83.

[9] Shell, SECOND SHELL OMEGA PROCESS PLANT STARTS UP IN SAUDI ARABIA, Press release, June 2nd, 2009. Sourced 6 Aug. 2019: https://webcache.googleusercontent.com/search?q=cache:pF0IRaMHkVIJ: https://www.shell.com/business-customers/chemicals/media-releases/2009-media-releases/pr-startup-second-shell-omega-process-plant.html+&cd=2&hl=en&ct=clnk&gl=nl

[10] Shell, "Shell starts-up world-scale monoethylene glycol plant in Singapore" (Press release). Shell Chemicals. 2009-11-17. Sourced 9 Aug 2019: https://www.shell.com/business-customers/chemicals/media-releases/192009-media-releases/shell-starts-meg-in-singapore.html

[11] Shell, Omega and Ethylene oxide/Ethylene Glycol Technology, Shell website, 2019. Sourced 9 Aug. 2019: https://www.shell.com/business-customers/chemicals/factsheets-speeches-and-articles/factsheets/omega.html

[12] Scholes, G, Integrating Sustainable Development into the Shell Chemicals Business, in Weijnen MPC, Herder, PN (Ed), Environmental Performance Indicators in the Process Design and Operation, EFCE Event 616, Delft University Press, 1999, 69–82.

[13] Albuloushi N, Algharaballi E. Examining the influence of the cultural aspect of uncertainty avoidance on supply chain coordination. Journal of Applied Business Research (JABR). 2014 Apr 24;30(3), 847–862.

# 16 Intensified ethanol production by cyclic distillation

## 16.1 Background information

Ethanol can be produced by fermentation of sugars by yeasts or via petrochemical processes. It is used in the food industry (recreational drug), for medical applications (antiseptic and disinfectant), as chemical solvent in manufacturing processes, and also as a renewable fuel (bioethanol).

The flowsheet of a typical ethanol production plant is shown in Figure 16.1 and consists of a beer column containing pre-concentrating ethanol (C1), a hydroselection column (C2), a rectification column (C3), a column for end-cleaning which produces high purity ethanol (C4), a column for concentrating impurities (C5), a fusel column (C6), and a methanol column (C7). The light (head), intermediate, and heavy (end) impurities are removed by distillation columns C2, C5, and C7. The main goals in the purification of ethanol are to obtain a product with a minimum impurities and to maximize the yield of products. Additional energy savings can be achieved by operating the

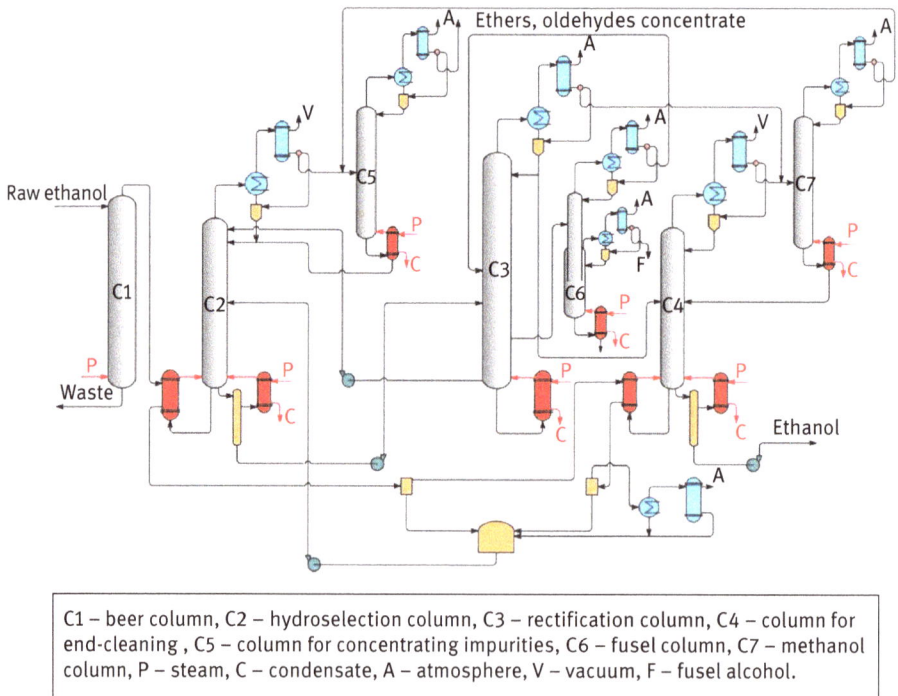

C1 – beer column, C2 – hydroselection column, C3 – rectification column, C4 – column for end-cleaning , C5 – column for concentrating impurities, C6 – fusel column, C7 – methanol column, P – steam, C – condensate, A – atmosphere, V – vacuum, F – fusel alcohol.

**Figure 16.1:** Flowsheet of an ethanol production plant.

https://doi.org/10.1515/9783110657357-016

columns at different pressures, as for example, the vapor from the beer column is used to heat up the hydroselection and the methanol columns [1].

This chapter aims to provide valuable academic and industrial insights into using cyclic distillation technology in the ethanol production with a focus on the equipment supplier perspective.

In addition, it provides information on how the technology was developed and what are the roles of the soft elements as explained in Chapters 3–5.

## 16.2 Discovery stage

Cyclic distillation is an intensified technology for fluid separations, which uses a different approach in contacting the liquid and vapor phases [2]. Unlike conventional operation, cyclic distillation uses separate phase movement that can be achieved with specific internals and a periodic operation mode [3]. One operating cycle consists of two key parts: a vapor flow period (when the thrust of rising vapor prevents liquid down flow) followed by a liquid flow period (when the liquid flows down the column, dropping by gravity, first to a lock chamber and then moving to the tray below) – see Figure 16.2 [4].

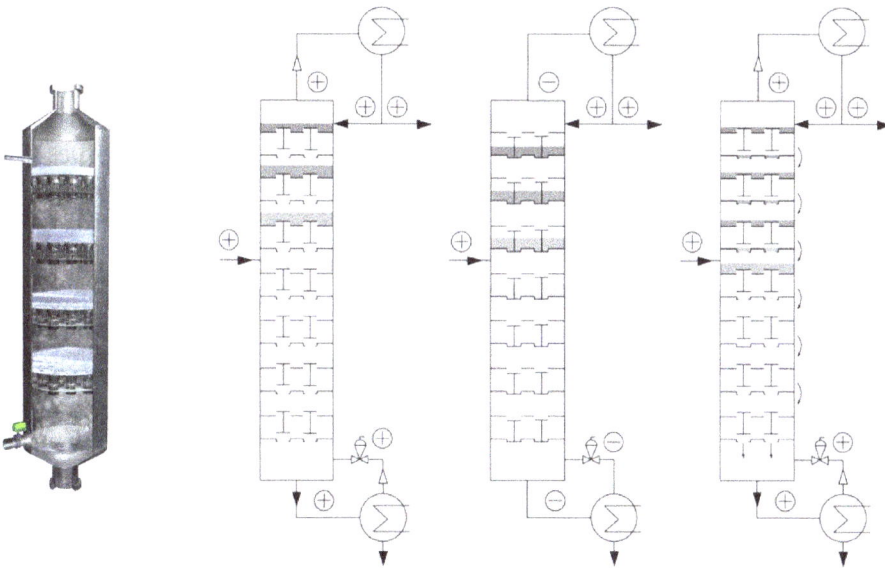

Figure 16.2: Cyclic distillation column vapor flow and half-liquid overflow periods.

Several papers and literatures on cyclic distillation are available, covering its discovery and history (from the late 1930s through various stages of renewed attention in the 1930s, 1960s, and 1980s, until the first industrial-scale applications after 2000), working principles, modeling and simulation [5], the simultaneous vs consecutive cycling operation mode [6], the perfect displacement model and working lines [2], mathematical modeling [7, 8], driving force based design [9], design and control [8, 10], impact of operating parameters on the performance [11], new tray designs [12], pilot-scale studies [13], industrial equipment and applications [14], and revamping of conventional columns [3].

## 16.3 Concept stage

A cyclic distillation column looks just like a regular tower from outside. However, the cross-section view inside a cyclic distillation column reveals the absence of down comers and the presence of different internals that allow an efficient separate phase movement in practical operation. An analogue exists in the cyclic distillation, where time is the independent variable while in conventional distillation, it is distance (). Conceptually, a cyclic distillation column resembles a series of batch distillations.

In the concept stage, MaletaCD designed and operated a pilot-scale process. The column for concentrating impurities is part of an industrial plant producing food grade alcohol (Lipnitsky alcohol plant, Ukraine). Figure 16.3 shows the flowsheet of the ethanol concentration section with the cyclic distillation column as the key operating unit [2]. The wastes from this column are 5% of the plant capacity. The column concentrating the impurities recycles the waste back to the plant. Remarkably, the wastes of this column are only 0.5% of the plant capacity and hence, an additional alcohol output (4.5% of the total plant capacity) is possible.

The hydroselection column and the column for concentrating impurities in the process are of the same type, performing the same task of removing the volatile components while sharing identical impurities, and working under the same conditions. The main difference between them lies in the fact that the concentration of impurities in the cyclic distillation column is much higher than in the hydroselection column. This makes it possible to increase the yield of the desired product. The efficiency of both columns is compared under conditions close to the ones determined by Fenske method. For both columns, the reflux ratio is about 50 mol/mol. Table 16.1 presents the concentration of impurities (ppm) in the liquid stream for the hydroselection column and also for the column concentrating the impurities [2].

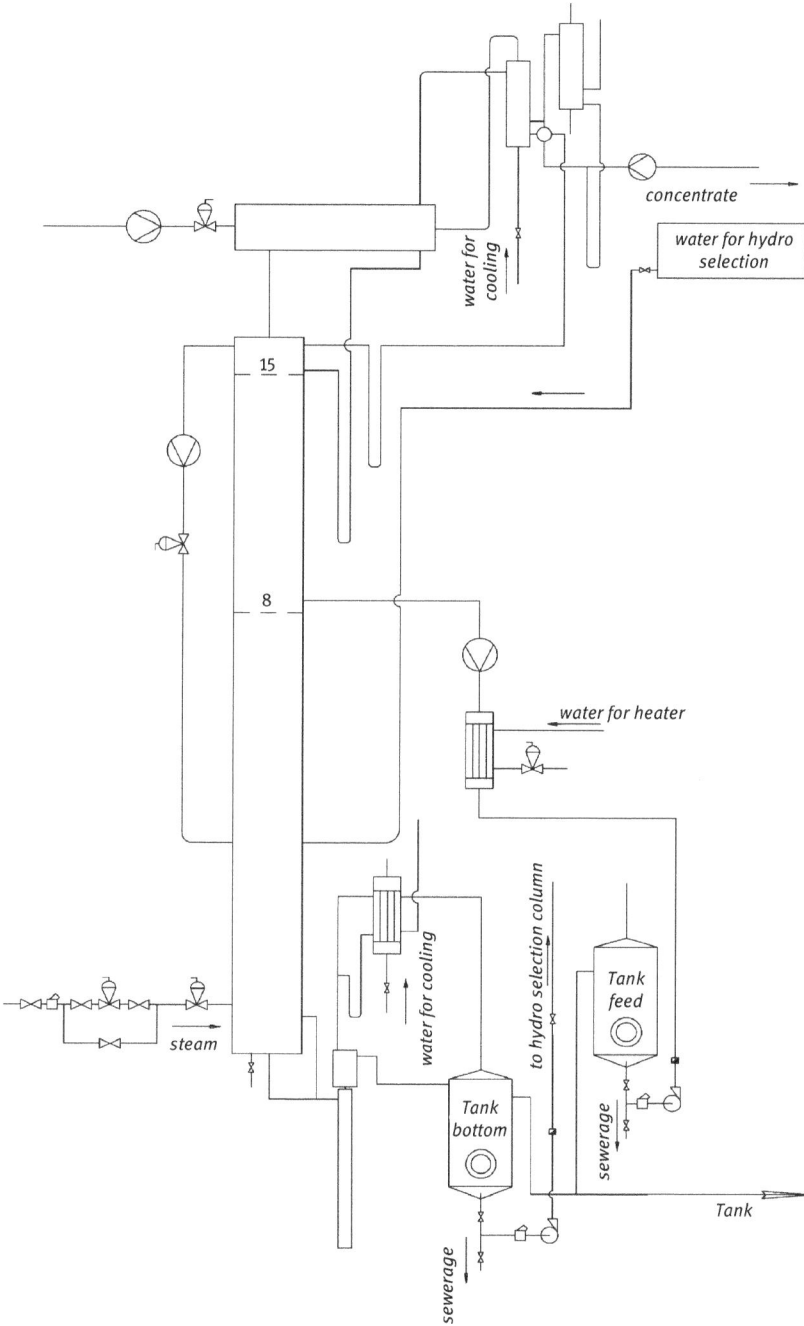

**Figure 16.3:** Cyclic distillation column for ethanol concentration.

**Table 16.1:** Impurity content (ppm) in feed and products for a capacity of 20,000 L/day.

| Component | Group | Hydroselection column (conventional operation) | | | Column concentrating the impurities, cyclic mode | | |
|---|---|---|---|---|---|---|---|
| | | D = 1,000 mm, 56 trays, W = 550 kW | | | D = 400 mm, 15 trays, W = 50 kW | | |
| | | Feed | Bottom | Top | Feed | Bottom | Top |
| Ethylic ether | Ether Aldehyde | – | – | 0.223 | 0.223 | – | 1.404 |
| Acetic aldehyde | Ester | 88.29 | 0.461 | 660.38 | 660.382 | 0.36 | 17,730 |
| Methyl acetate | Ketone | 6.597 | – | 103.72 | 103.727 | – | 2,394 |
| Acetone | Ester | 0.727 | – | 7.3 | 7.3 | – | 192.2 |
| Ethyl acetate | Alcohol | 261.5 | – | 3,806 | 3806.69 | – | 94,527 |
| Methanol | Ketone | 0.038 | 0.005 | 0.15 | 0.150 | 0.009 | 1.94 |
| 2-Butanone | Alcohol | – | 1.339 | 5.38 | 5.383 | – | 115.37 |
| 2-Propanol | Ether | 5.77 | 0.393 | 75.17 | 75.178 | 2.93 | 1,796 |
| Isobutyl acetate | Alcohol | 2.00 | 2,289 | 57.85 | 57.856 | 0.17 | 1,465 |
| 1-Propanol | Aldehyde | 13,762 | 0.075 | 83.76 | 83.763 | 80.04 | 1,192 |
| Croton aldehyde | Alcohol | 0.655 | 2,661 | 5.22 | 5.226 | 0.201 | 143.6 |
| Isobutyl alcohol | Esther | 23,476 | – | 84.64 | 84.642 | 139.86 | 57.6 |
| Isoamyl acetate | Alcohol | – | 18.72 | 57.17 | 57.175 | – | 1,272 |
| 1-Butanol | Alcohol | 130.4 | 4,295 | – | – | 0.744 | 6.62 |
| Isoamyl alcohol | Alcohol | 42,982 | 1.193 | 3.43 | 3.431 | 205.64 | 78.49 |
| 1-Pentanol | Alcohol | 11.82 | 1.936 | – | – | – | 3.64 |
| 1-Hexanol | Arom. aldeh. | 20.20 | 0.809 | – | – | 0.142 | 0.26 |
| Benzaldehyde | Arom. alcohol | 6.42 | 42.43 | – | – | 0.309 | – |
| 2- Phenyl ethanol | | 518.5 | | – | – | 1.088 | – |

## 16.4 Feasibility stage

The first step in the development of the cyclic distillation technology was a scientific activity at the National University of Food Technology in Ukraine. As a result of the work, information was obtained that allowed at all stages, for all columns, to intensify the process of purification of ethanol food grade. Column C1 was selected for the physical modeling, and then successful tests were carried out on the practical results of the new technology. However, in order to minimize the risks associated with the industrial implementation, it was decided to first replace the industrial column C5 used for concentrating impurities. Another reason was that the industry was interested in improving product quality and minimizing production waste. Improving the quality of food grade ethanol is associated with the processing of the main contaminated streams by column C5 and increasing the yield due to working at higher concentration of impurities.

## 16.5  Development stage

MaletaCD is an engineering company developing the cyclic distillation technology. Given the wide range of distillation and sorption processes using columns, a key activity is aimed at adapting cyclic distillation technology to the many areas of application, from ethanol purification to isotope separation. But distillation remains a fairly conservative area as it much associated with significant investment costs. As the practical implementation practice has shown, the customer is ready to take the risks of the new technology only in extreme cases, when all traditional technology possibilities are exhausted. This was also the story in case of MaletaCD. Having exhausted all possible options for improving the quality of food grade ethanol, the Lipnitsky distillery took on the risks of using cyclic distillation technology in an industrial environment. The plant's capacity was 20,000 L of food ethanol per day. As a result of the industrial tests, additional refinement of special trays was required both in the materials of the elements and in the geometric ratio of the parameters of the mass contact devices, which were further developed and then patented by MaletaCD.

## 16.6  Engineering procurement and construction stage

The theoretical work on cyclic distillation was sufficiently developed at the initial stage of MaletaCD's activity. The main problem of industrial implementation was that the existing trays could not take advantage of the new technology. The main condition for realizing the advantages of cyclic distillation is the lack of mixing of liquid on adjacent trays during the overflow of liquid from one tray to another. MaletaCD proposed the concept of new mass transfer contact devices which uses a lock chamber under the tray. Thus, the movement of fluid along the column was carried out according to the scheme: bubble tray → sluice chamber → empty bubble tray below. The liquid flow time on the trays is standard for all such trays, about 2–3 s. The control effect on the fluid flow is provided by steam which has a sufficient flow rate to keep the liquid on trays. Then, the steam supply to the column is stopped for 2–3 s to allow the simultaneous liquid drain from one tray to another. Automatic "butterfly" valves are used for steam flow control. This method of cyclic distillation allows for the separate movement of phases in a wide range of liquid and vapor loads. With a constant amount of liquid on a tray, the liquid loads are determined by the frequency of the cycles. The steam velocity in the column for these devices is always higher than critical and therefore, its increase is limited only by the transfer of liquid to the tray above.

Table 16.2 provides a list of industrial implementations of cyclic distillation by MaletaCD, in the food industry for the production of ethanol (in Eastern Europe). More industrial implementations are currently under development and/or construction.

**Table 16.2:** Industrial implementations of cyclic distillation by MaletaCD in ethanol production.

| No. | Name of the facility | Capacity of the plant (L/day) | Application (column) | Column diameter (mm) | Number of trays (psc.) | Feed to column (m$^3$/h) |
|-----|---------------------|------------------------------|---------------------|---------------------|-----------------------|-------------------------|
| 1 | Lipnitsky ethanol factory (Ukraine) | 20,000 | Impurity concentration column | 400 | 15 | 1.2 |
| 2 | Kosare ethanol factory (Ukraine) | 50,000 | Impurity concentration column | 650 | 15 | 3.5 |
| 3 | Zalozetskiy ethanol factory (Ukraine) | 30,000 | Impurity concentration column | 560 | 15 | 2.0 |
| 4 | Nemirovsky ethanol factory (Ukraine) | 30,000 | Impurity concentration column | 560 | 15 | 2.0 |
| 5 | Chervonoslobodskoy ethanol factory (Ukraine) | 30,000 | Impurity concentration column | 560 | 15 | 2.0 |
| | | | Hydroselective column | 1,200 | 21 | 4.2 |
| | | | Regeneration column | 800 | 21 | 2.2 |
| | | | Secondary beer column | 700 | 15 | 2.5 |
| 6 | Malinovshchinsky ethanol factory (Republic of Belarus) | 30,000 | Impurity concentration column | 560 | 15 | 2.0 |

## 16.7 Start-up stage

The launch of the first industrial column in Lipnitsky took place just before the New Year and the column reached a (pseudo-)steady state within hours. However, the tower stopped 2 weeks later. Armor-piercing organic glass was used as the material for the valves. When disassembling the column, it was found that the valves, when exposed to high concentrations of impurities, turned into wet cardboard paper and fell apart. The valve material was then replaced by stainless steel which required additional coordination of the geometric parameters of the mass transfer contact devices.

## 16.8 Evaluation

Cyclic distillation has enormous potential in various industries, especially since it is associated with obtaining high purity products. This is the case for the food industry, chemistry, petrochemistry, pharmaceuticals, organic synthesis, and isotope separation.

However, given the conservatism of the distillation technology users and suppliers, this process will require a lot of time and significant investment. Customers are usually interested in no risk in the implementation of this technology, over and above the benefits proven in implemented projects. A variety of technological schemes and products require the testing of new processes in a pilot unit for every industrial sector.

### 16.8.1 Sustainable development goals

The enhancement of separation efficiency realized in cyclic distillation resulted in a number of significant advantages over traditional processes. The energy usage is reduced by 20–30% leading to a decrease in the consumption of utilities, primary energy, and water resources, while also reducing greenhouse gas emissions associated with the energy usage. Increased separation efficiency means that the cyclic distillation columns have a 1.5–2 times lower height compared to classic technology. This reduces the size of the equipment and investments in the column (shell & internals), heat exchangers, and the supporting building structure. An increase in the separation efficiency of components invariably also entails an increase in product quality which can guarantee a higher market price of the product. In summary, the cyclic distillation technology especially contributes to the SDG 12 (Ensure sustainable consumption and production patterns) and also to SDG 7 (Ensure access to affordable, reliable, sustainable and modern energy for all) by increasing the economic growth of a country.

### 16.8.2 PI concepts

Cyclic distillation is an excellent example of process intensification in the temporal domain. By using an alternative operating mode with separate phase movement, the energy and separation efficiency are increased, leading to lower capital cost and lower energy requirements. So, it is clearly a prime example of an intensified process with economic and sustainability benefits.

### 16.8.3 VIB perspectives

The **V**alues of engineers, **I**nterests of stakeholders, and **B**eliefs of society – known as the VIB perspectives model in this book – results in guiding principles for each innovation stage. The values of the engineers refer to the values that motivated the chemical and mechanical engineers to develop the intensified process. In case of cyclic distillation, these *values of engineers* were related to:

- Production: increase in end product output per unit of raw materials, reduction of energy costs per unit of product, higher efficiency, less equipment required, and decrease in investment costs.
- Consumer: higher purity of the end product, reduction of harmful impurities in the end product (e.g., esters and methanol), and improvement of the organoleptic indicators.
- Sustainability: lower carbon footprint (less pollution, lower $CO_2$ emissions) by reduced energy usage, reduced consumption of water resources, and less materials used in the new equipment.

*The interests of stakeholders* refer to the interests of external stakeholders, the most common stakeholders being suppliers of material and technology, customers, local and national authorities, shareholders, knowledge institutes, the local community, social action groups, and so on. In the case of the first cyclic distillation implementation, the stakeholders of the client (Lipnitsky) were the most important. The owner of the plant was most interested because of solving a production problem while getting financial gains along with some sustainability benefits that painted a positive picture in the community and with the authorities. However, the technical staff of the plant was more reluctant to change as they perceived it as getting new headaches in the implementation process (e.g., additional work, risks of mistakes, significant mental effort). In hindsight, perhaps it would have helped the whole process if the end user technical staff was involved in an earlier phase to get, in advance, more hands-on experience and understanding of the cyclic distillation technology.

*The beliefs in society* refer to the beliefs in the country in which the technology is implemented. In the case of cyclic distillation, the innovation was implemented in a production plant in Eastern Europe where the alcohol consumption is very popular and this trend is expected to remain in the future. The main problem is to reduce the consequences of alcohol usage and if the product is of high quality, then some of the negative consequences of its use will be reduced. To be fair, the bulk of the alcohol consumers are indifferent to what method is used to make the product. However, there is a societal support that welcomes innovation if it leads to key benefits for the regular consumer.

### 16.8.4 Multidisciplinary cooperation

Multidisciplinary cooperation is very beneficial for rapid innovation and implementation of this new process technology. In the case of developing the cyclic distillation technology, there was a very good collaboration between (chemical and mechanical) engineers and chemists from industry (end-users and technology suppliers) and academia.

In our approach, the multidisciplinary cooperation focused especially on the co-operation between Research & Development (R&D), Operations (OPS), and Marketing & Sales (M&S). The R&D group of MaletaCD, the developer of the cyclic distillation technology, was the driver and acted as the technology push. Marketing & Sales (M&S) of the end user (Lipnitsky) was on the business pull side as they really wanted to improve the quality of the alcohol produced significantly, which was rather poor. Operations (OPS) of both the equipment supplier and industrial end user contributed to taking the cyclic distillation technology from the pilot scale to the implementation at large scale.

It should be noted that the traditional distillation columns with cap trays did not work properly and this was the business urgency that required immediate action. Fortunately, at that time, MaletaCD was testing the pilot plant setup of its cyclic distillation technology. Some technical representatives of the Lipnitsky plant attended these pilot-scale tests and got acquainted with the cyclic operation of the column. Convinced of the reliable operation and promised benefits, they ventured to install a cyclic distillation column into their factory and the rest is history.

## 16.9 Learning points

It is clear from the case description that *SDGs* can play an important role in increasing the market penetration of cyclic distillation in food and other process industry branches.

Cyclic distillation is a good example of applying the PI *temporal domain* in design.

To obtain learning points for the use of *VIB perspectives* and, in particular, to obtain more background on the education and drive of the engineers, Vladimir Maleta interviewed Tony Kiss to push the cyclic distillation technology to commercial implementation. Tony in turn asked Vladimir for additional background. The interview results are found in the text below.

As a child, Vladimir was an avid reader and was driven by curiosity, especially in science. He graduated from the National University of Food Technologies in Ukraine (and supported his children to do the same). In early 1980s he came across cyclic distillation. There were many articles and publications and great prospects for its use. His understanding of the topic was that traditional trays cannot realize all the benefits of the new process (rightfully so), as they could not ensure separate phase movement. So, his task was clear as he embarked on creating a tray design that eliminates fluid mixing during fluid overflow. He went to the technical library and spent a lot of time studying various designs of contact devices. Once, in the central park in Kiev, he realized the engineering idea of implementing this in practice: the bubbler of the tray – the airlock – the bubbler of the lower tray, etc. and also the way to implement this. And this is how the new Maleta internals were born and he remained committed to get this technology implemented industrially.

Our collaboration is based primarily on mutual deep interest in process intensification. The interest is deep (not formal due to professional duty). Both Vladimir and I have a common interest in intensified distillation technologies (I even wrote a book on this in 2013 [14]). As for Vladimir, he is working on taking cyclic distillation to the next level, for example, by combining it with catalytic distillation or DWC technology.

I met Vladimir at the CHISA-2010 conference, at the Distillation & Absorption section (where I gave a presentation), and we had some interesting and lengthy discussions about advanced distillation technologies. I was also impressed by the fact that he had, in his bag, the Maleta internals for cyclic distillation and he showed me exactly how they work. We clicked in terms of professional and scientific interests, and ever since we have worked together and published several joint articles.

My PhD was in the Process Systems Engineering area, dealing with the design and control of chemical processes involving recycles by non-linear analysis. This basically covers the large majority of processes in the chemical industry based on reactor-separator-recycle structures. As distillation is the most used separation method, it was also part of my focus along with improving energy efficiency, cost-effectiveness, and other performance indicators. This naturally led to extensive work on PI technologies such as reactive distillation, dividing-wall column, cyclic distillation, etc. (www.tonykiss.com).

As for Vladimir, he associates himself with the cyclic distillation technology. This being his personal / professional focus, he wants to keep having industrial successes and more satisfied clients. Along with the acquisition of new knowledge comes personal development and this stimulates and gives him energy for moving forward further.

Vladimir persisted to implement cyclic distillation in the food industry against the conservative attitude. He found that it is important to determine the "business pain" that requires urgent action (e.g., loss of production or insufficient product quality) and pair it with the promising gain of his new technology that is able to address these food branches issues. Once the first installation of a technology proved to be successful, other industrial implementations followed.

So, the learning point is that the *value of engineers*, shown in the personal engineering drive of Vladimir and Tony played an overriding role in the successful implementation of the Maleta cyclic distillation technology.

*Multidisciplinary cooperation*, in combination with taking the *interest of the stakeholder*, the client Lipnitsky, into account, also played a role. Those two elements are revealed in the intimate cooperation between Maleta and Lipnitsky. Both belonging to the same culture of Ukraine probably also helped in this respect.

# References

[1]    Maleta, VN, Bedryk, O, Shevchenko, A, and Kiss, AA, Pilot-scale experimental studies on
       ethanol purification by cyclic stripping, AIChE Journal, 2019, 65, e16673.
[2]    Maleta, VN, Kiss, AA, Taran, VM, and Maleta, BV, Understanding process intensification in
       cyclic distillation systems, Chemical Engineering and Processing: Process Intensification,
       2011, 50, 655–664.
[3]    Kiss, AA, and Bildea, CS, Revive your columns with cyclic distillation, Chemical Engineering
       Progress, 2015, 111(12), 21–27.
[4]    Kiss, AA, and Maleta, VN, Cyclic distillation technology – A new challenger in fluid
       separations, Chemical Engineering Transactions, 2018, 69, 823–828.
[5]    Bildea, CS, Patrut, C, Jorgensen, SB, Abildskov, J, and Kiss, AA, Cyclic distillation
       technology – A mini-review, Journal of Chemical Technology and Biotechnology, 2016, 91,
       1215–1223.
[6]    Toftegård, B, Clausen, CH, Jorgensen, SB, and Abildskov, J, New realization of periodic cycled
       separation, Industrial & Engineering Chemistry Research, 2016, 55(6), 1720–1730.
[7]    Krivosheev, VP, and Anufriev, AV, Mathematical modeling of the cyclic distillation of binary
       mixtures with a continuous supply of streams to the column, Theoretical Foundations of
       Chemical Engineering, 2018, 52(3), 307–315.
[8]    Patrut, C, Bildea, CS, Lita, I, and Kiss, AA, Cyclic distillation – Design, control and
       applications, Separation & Purification Technology, 2014, 125, 326–336.
[9]    Nielsen, RF, Huusom, JK, and Abildskov, J, Driving force based design of cyclic distillation,
       Industrial & Engineering Chemistry Research, 2017, 56(38), 10833–10844.
[10]   Andersen, BA, Nielsen, RF, Udugama, IA, Papadakis, E, Gernaey, KV, Huusom, JK, Mansouri,
       SS, and Abildskov, J, Integrated process design and control of cyclic distillation columns,
       IFAC-PapersOnLine, 2018, 51(18), 542–547.
[11]   Buetehorn, S, Paschold, J, Andres, T, Shilkin, A, and Knoesche, C, Impact of the duration of
       the vapor flow period on the performance of a cyclic distillation, ChemieIngenieurTechnik,
       2015, 87, 1070–1070.
[12]   Maleta, BV, and Maleta, O, Mass exchange contact device., US Patent 8,158,073, April 17,
       2012.
[13]   Maleta, BV, Shevchenko, A, Bedryk, O, and Kiss, AA, Pilot-scale studies of process
       intensification by cyclic distillation, AIChE Journal, 2015, 61, 2581–2591.
[14]   Kiss, AA, Advanced distillation technologies – Design, control and applications, JohnWiley&
       Sons, Inc, Chichester,UK, 2013.

# 17 Pharmaceuticals case

## 17.1 Background information

### 17.1.1 Background pharmaceutical industries

In pharmaceutical industries, active pharmaceutical ingredients (APIs) have a central role in the drug manufacturing chain. The industry itself is highly regulated by authorities of different countries where drugs and APIs are manufactured and exported. In European Union, the European Medicines Agency (EMA) [1] was founded to harmonize actions of national agencies in the evaluation and supervision of medicinal products. The corresponding agencies are in US Food and Drug Administration (FDA) [2] and Pharmaceuticals and Medical Devices Agency (PMDA) [3] in Japan. They have responsibilities to draw up Acts for supervising drug manufacturing. The International Council for Harmonisation of Technical Requirements for Pharmaceuticals for Human Use (ICH) [4] harmonizes recommendations for pharmaceutical industry and it can be considered as an umbrella of Pharma Industries. Additionally, regulations by other national bodies concerning safety, health, and environment are strictly followed by responsible manufacturing companies.

Pharmaceutical industries rely heavily on traditional manufacturing methods that have been used for decades, independent of the company location, globally. One of the reasons is that new drugs are urgently needed to be introduced to the market and this is possible to be achieved with traditional process solutions. Another reason may be that the regulations are easier to fulfill based on traditional production solutions, that is, by imitating old technologies. The core of API processing is based on batch production which consists of pretreatment of raw materials, reactor operation, product recovery, and further work-up processing (see Figure 17.1). There are typically 4–8 reaction syntheses steps and at least 2–3 intermediate products are isolated before the final API is ready. The majority of APIs are in crystalline form when they are used as raw materials for drugs. Batch processing is a typical technology when relatively low production capacity (<100–200 t/year) is concerned, and it is also practical when several different products are produced in multipurpose equipments.

### 17.1.2 Process intensification limitations in pharma industries

The process Intensification paradigm in Pharma Industry has significant limitations compared to the statement of ERPI [5]: "A set of often radically innovative principles ('paradigm shift') in process and equipment design, which can bring significant (more than factor 2) benefits in terms of process and chain efficiency, capital and operating expenses." Especially, equipment design is ruled out from practical

https://doi.org/10.1515/9783110657357-017

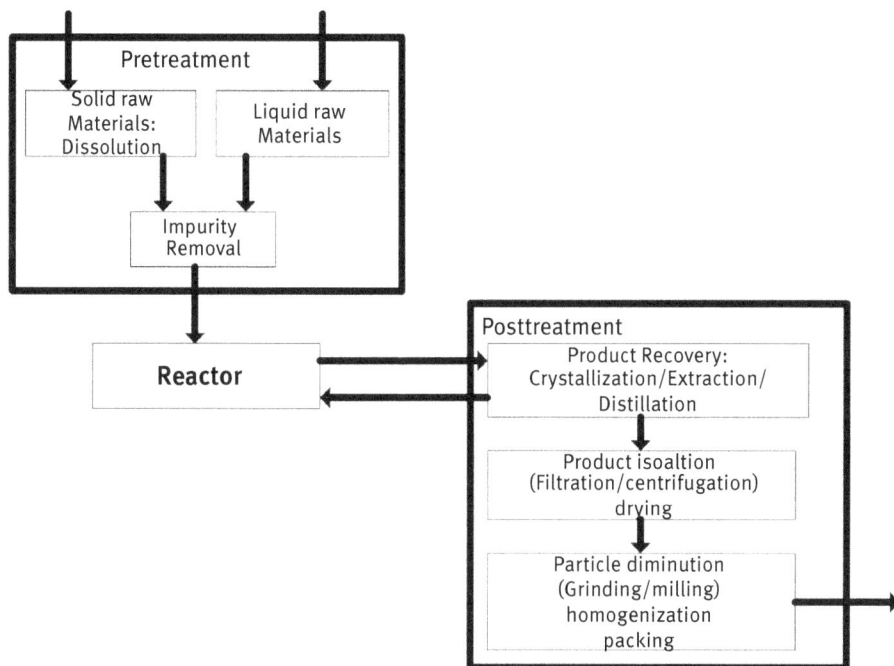

**Figure 17.1:** Typical batch production scheme in API production.

work due to the above mentioned reasons but several radical functional improvements [6] can be carried out in spite of those limitations. Significant process improvements are also challenging for API products because the process documentation is filed for the use of Authorities and in this sector, there is a lack of example cases in continuous process solutions. The process changes are classified by the authorities into different categories, from major to minor changes, according to ICH guidelines [7] although quality by design [8–10] approach allows some freedom of action. However, radical innovations for improving current processes should possess considerable benefits for Pharma companies in terms of profitability increase or when legislative compulsion drives continuation of production. Driving legislation can be, for example, environmental or process safety issues. As a rule of thumb, the manufacturing process in API filing material should be described only in adequate accuracy by pointing out the essential facts related to product quality. If it is too detailed a documentation, future changes to improve manufacturing process would take more resources and time to get approval by FDA, EMA, and so on.

One of the major principles in API product filing is to describe the manufacturing process as close as possible to the final molecule structure. It means, to select raw materials for the process, the description is to include the last chemical reactions where API molecule is created. However, the raw materials quality should be described at an adequate level for a viable control of final product quality. This strategy thus allows to improve earlier manufacturing steps with more freedom.

Another principle is to describe the equipment in the manufacturing process as broad as possible so that equivalent equipment selection is possible, keeping in mind not to decrease the final product quality. For example, the reactor volumes of equivalent equipment should be of the same size, otherwise the scale-up effect should be verified for preserving the product quality, and often the change can be categorized as a major change by the authorities.

When process changes are made, the new improved manufacturing process should always be validated even in case of raw material, to maintain good manufacturing practices (GMP), high level of quality control (QC), and for quality assurance (QA).

In the remaining sections of this chapter, the process intensification dealkylation case is described in detail. Appendix I contains an additional pharmaceuticals case where evaporation and cooling crystallization steps are combined.

### 17.1.3 Process intensification case dealkylation

The purpose of this case is to show the basic principles and ideas that can be used to intensify chemical processes which are considered relevant in the context of Process Intensification. The process solution presented here is meant to be useful for adopting similar process improvements to different pharmaceutical industrial processes.

The example case also takes into account an evaluation based on values of engineers, interests of stakeholders, and beliefs of society. The example case is, furthermore, representative for innovations without making any improvements in existing process equipment selection. The example case is related to significant increases in production capacities and reduction of production times. The chemistry of the PI case is substituting the environmentally difficult dealkylating metal chloride with dealkylating by $H_2SO_4$.

The project target is to decrease the production time and substitute an environmentally toxic reagent by a non-toxic agent in a dealkylation process.

The reaction synthesis starting material is Ph-O-R and the intermediate product is Ph-OH. Metal chloride, as a reagent, is substituted with concentrated sulfuric acid (min 97 wt%). Metal chloride based process produces an environmentally difficult reaction mixture as a waste stream which can only be destroyed by a high temperature combustion. The sulfuric acid is a strongly dealkylating reagent which forms large amounts of cyclic and aromatic tars if the reaction is not carefully designed and if it does not contain a water quenching stage. Tars have a considerable negative effect on product color and on product recovery by centrifugation. The organic sulfates formed in dealkylation are neutralized by acid hydrolysis before discharging industrial waste water treatment plant.

The core radical innovative step in the process design is the controlled crystallization of dealkylated product from the reaction mixture, to get sufficient settling of formed particles. This is done by adding Ostwald ripening (temperature swing) for increasing the crystal settling velocity from less than 100 mm/h to 500 mm/h, based on experiments. The next PI step is to recover the crystalline product by decanting the reaction mixture using dip pipe. In the water addition stage, after the first decantation, water is added to reduce impurities in the reaction mixture by dilution and increase the settling velocity by decreasing the fluid viscosity. Decantation is to avoid a centrifugation step and to directly purify the crystallization without changing reactors. The tars are removed from the reaction mixture by dissolving the formed crystals in a heating step, followed by filtrating the solution using normal active carbon.

The product is recovered using cooling crystallization, and by centrifugation separation of crystal product from the mother liquor. Finally, the intermediate product is dried. The total process time is less than 50 h using this sulfuric acid dealkylation, decantation of reaction product, and including a final crystallization purification step, directly from decanted residue after active coal filtration. Block diagrams of metal chloride and sulfuric acid based dealkylation processes are presented in Figure 17.2.

In the metal chloride dealkylation, the reaction product is crystallized and recovered by centrifugation. The recovered crystal product is dissolved in a separate reactor before active coal hot filtration and then recrystallized in an additional reactor. The total time, including final product centrifugation and drying, is over 70 h. Additionally, the sulfuric acid dealkylation increased the profitability due to simple waste treatment, compared to metal chloride-based waste treatment by high temperature combustion technology.

The differences between the old and the new intensified process are summarized in Table 17.1. Also the differences in values of engineers and interests of stakeholders (environmental permits from authorities, QA&QC, and eventually company ownership) are summed up.

## 17.2 Discovery stage in dealkylation process

The metal chloride substitution with sulfuric acid was first tested in a lab in the beginning of the 1990s, but the product yield and solid product recovery was not impressive. Sulfuric acid route was rejected. Twenty years later, this synthesis route was tested and it was found that the reaction could be optimized. The reaction was first optimized with oleum (fuming sulfuric acid) due to minimization of water. The presence of water during the reaction step was partly related to formation of tars, which was due to polymerization of dealkylated product. It was soon found out that concentrated sulfuric acid with careful design of the temperature profile also worked well. Eventually, Ostwald ripening was discovered and the overall process

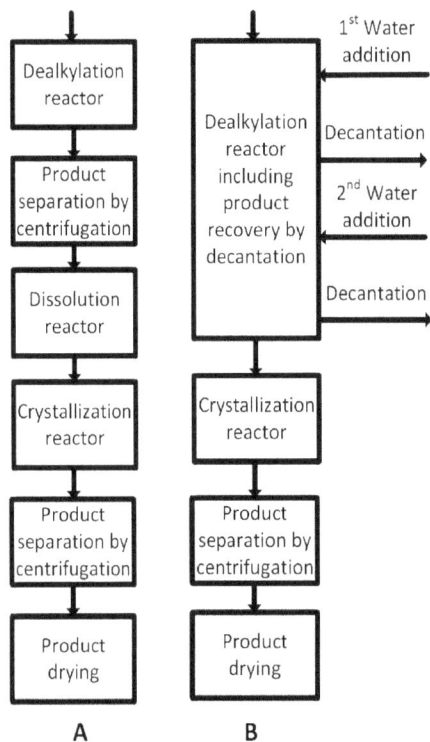

**Figure 17.2:** Block diagram unit operations dealkylation process. (A) Old process, (B) intensified process.

was found successful in the lab. Filtration and settling times were documented and at this point, crystal properties also indicated that the product can be separated from the reaction mixture by centrifugation. Later, recrystallization was done by dissolving the crystals and using hot filtration.

The discovery stage is described in Appendix 1 for an additional pharmaceuticals case.

## 17.3 Concept stage in dealkylation process

The concept was first verified in a lab-scale setup in which the factory-scale batch processing steps were simulated and then bench-scale batches were performed similar to an industrial operation. At this point, the acid hydrolysis was also developed in lab scale so that wastewater treatment could be tested. Product recovery from the reaction mixture was obtained by filtration.

The process profitability was initially evaluated based on these lab-scale experiments.

**Table 17.1:** Comparison between old process and process-intensified process: production time decrease and substitution of environmentally toxic reagent in dealkylation process.

| | Old process | New process |
|---|---|---|
| Dealkylated reagent | $MeCl_2$ | $H_2SO_4$ |
| Solvent | Organic | Min. 97 wt% $H_2SO_4$ |
| Waste stream | Metal-based reaction mixture | Alkyl sulfates containing water solution |
| Product recovery from reaction mixture | Cooling crystallization from organic solvent | Semibatch precipitation from NaOH–water (acts also as a quenching reagent) |
| Work-up | Centrifugation, filter cake washing | I reaction mixture decantation, I water addition, II diluted reaction mixture decantation, II water addiction |
| Purification step | Solvent exchange and charring crude product, dissolving crude product, active coal filtration, cooling crystallization | Dissolving product into water, active coal filtration, cooling crystallization |
| Work-up | Centrifugation, filter cake washing, drying | Centrifugation, filter cake washing, drying |
| Processing time | >70 h | <50 h |
| Number of equipments | 6: 3 reactors, 2 centrifuges, 1 dryer | 4:2 reactors, 1 centrifuge, 1 dryer |
| Waste treatment | High temperature combustion, use of contract incinerator company, organic solvent regeneration | Acid hydrolysis and transfer to in-house wastewater plant |
| PI concept | Multifunctions in reactor 1: (1) organic synthesis, (2) simultaneous quenching and precipitation, (3) decantation work-up to eliminate the use of additional reactor and further work-up steps. Functions in reactor 2: (4) reactor change and impurity removal using hot filtration, (5) cooling crystallization | |
| Value of engineers | Idea discovery by conducting experimental plan, concept and feasibility tested by simulating industrial-scale steps at lab, development and implementation included by scaling-up and process validation, Only dip pipe acquisition and assembly. | |
| Interest of Stakeholders | Owner's perspective: Intermediate product price reduction. Customer perspective maintaining same product (impurity) specifications. QA/QC department: maintaining same product (impurity) specifications. Environmental authorities: no additional reporting needed due to process change. | |

## 17.4 Feasibility stage in alkylation process

In-house wastewater treatment, substantially faster process, and more intensive equipment use were the key factors for profitability improvements. This process intensification was part of a larger project where all production syntheses were re-evaluated and the economic potential was determined. The preliminary profitability figures of this process change was estimated to be around 2 M€/y. The decision was taken to continue the project and industrialize the new sulfuric acid synthesis because of the profitability increase.

## 17.5 Development stage in dealkylation process

Further development to industrial scale is typically done by designing a test batch followed by experimental validation batches. First, it was attempted to recover the crystalline product using a centrifuge to obtain a dry product but the centrifugation time appeared to be too long. In a discussion with the plant personnel about the problem, they proposed decanting. Laboratory experiments then showed that the settling velocity of crystals could be increased from less than 100 mm/h to 500 mm/h. So, it was decided to decant in two phases where the reaction mixture was diluted with water additions. There was no need for additional investment or for major construction because the dip pipes for this decanting were available in plant storage.

## 17.6 Implementation stage in dealkylation process

No formal EPC stage was present in this dealkylation process. The technology transfer to industrial plant was considered to be smooth without changing any process parameters, compared to bench-scale operation. Also, no procurement or equipment construction was needed.

The start-up for the commercial-scale production was in the existing equipment. Even this start-up is not marked as a separate stage. The first commercial-scale production is seen as the last part of the development.

In the second test batch and in three validation batches, decantation was used and later, product crystals dissolution could be done in the same reactor as dealkylation reaction was performed. Hot filtration was normally continued after dissolving of crystals and product recrystallization was done using cooling mode. Purified crystals were centrifuged and dried. The test batch and three validation batches were used for process validation documentation. The quality of the validation batch was within specification limits and incremental changes, either in particle quality properties or in impurity quality parameters, were not observed.

Development was carried out according to company's standard operating procedure (SOP) instructions which dictate development stages and reporting. The process times were evaluated based on test batch and validation batch times, which showed clear improvement when comparing the old process.

The advantage was, of course, reduction of reactors when product was not needed to take out from reactor.

## 17.7 Evaluation

### 17.7.1 Sustainable development goals

At the time of this project, the SDGs were not well known and there was no discussion in the company. So little specifics can be said on this subject.

### 17.7.2 PI concepts

The PI concept applied is summarized in Table 17.1. The main learning point was that reduction of reactor time using decantation also reduced the total processing time. Additionally, acid hydrolysis treatment for the reaction mixture waste led to a clear improvement in profitability.

The main PI domain applied is the plant-wide spatial domain in which the reaction chemistry is changed. Downstream processing is reduced not only in size and productivity, but efficiency is also improved. Pharmaceutical ingredients are substantially more expensive than bulk chemicals, where PI has been traditionally implemented. The significance of the process improvements related to product yield on starting materials can therefore have a big economic impact.

Generally, process intensification project in pharma involved understanding quality systems and chemistry. They are related to deep understanding of thermodynamics, reaction kinetics, and chemical synthesis mechanisms in order to do risk assessments and to convince the regulators, as discussed in some detail in Section 17.7.4.

### 17.7.3 VIB perspectives

The main element of the VIB perspectives relevant to this pharma case is the interest of stakeholders. As is well known, process intensification is about significant process improvements. This topic, in pharmaceutical development, is also deeply related to quality and safety. Initiative to process improvements starts from stakeholders' interests who are the process owners with profitability and sales volumes as interests. The authorities have legislation issues with health, safety, and environmental as interests. Customers have interests in product safety.

Every process change in pharma must be evaluated based on quality and safety. The tools are impact and risk assessment. The regulatory impact assessment means comparing the existing process and its description to intended changes. This part is carried out in cooperation with R&D, plant manufacturing personnel, regulatory affairs, and quality control personnel. Risk assessment is a continuous process during the whole process improvement project. It is carried out in the beginning of the project and during the transfer from laboratory to production site. Manufacturing personnel from the site, R&D, QA, and regulatory personnel are involved in risk identification, analysis, and in the evaluation steps. The limitations to significant improvements in Pharma thus come from regulations.

### 17.7.4 Multidisciplinary cooperation

As a process chemist, collaboration involved responsible personnel from factory, additionally including the HSE and QA experts. In the beginning of the PI project, attention was paid to environmental issues by HSE and the responsible factory personnel. Especially, acid hydrolysis was implemented at the factory plant site for neutralizing toxic waste before waste stream was discharged to the waste water treatment plant. Cooperation was intensified with the factory personnel when centrifugation was substituted with decantation for the preparation of the second test batch. Dip pipe installation to dealkylation reactor required careful design for efficient decantation mixture removal. After the process operator training for the new process, small changes for the process instructions were made during the execution of the second test batch. The successful cooperation of the factory plant personnel eventually enabled to industrialize the new process.

Appendix 1 describes a second pharma case with a similar multidisciplinary cooperation.

## 17.8 Learning points

The learning point from this case for the process intensification concepts is that process intensification can also be obtained in existing process equipment by changing the reaction chemistry. Applying the plant wide space domain method, less process steps are needed, leading to a shorter time.

The learning point for the stakeholder interests is that the regulatory authority is a very important stakeholder and that clear communication with that stakeholder is a key success factor for innovation projects in pharma industries.

The learning point on multidisciplinary cooperation is that many disciplines are involved. Interaction between plant personnel and the research process chemist plays a pivotal role in this multidisciplinary cooperation.

## References

[1]   European Medicines Agency, sourced 26-2-2020, https://www.ema.europa.eu/en
[2]   U.S. Food and Drug Administration, sourced, 26-2-2020, https://www.fda.gov/about-fda
[3]   Pharmaceuticals and Medical Devices Agency, sourced 26-2020, https://www.pmda.go.jp/english/about-pmda/index.html
[4]   The International Council for Harmonisation of Technical Requirements for Pharmaceuticals for Human Use, sourced 26-2-2020, https://www.ich.org/
[5]   European Roadmap for Process Intensification (ERPI). (2007), sourced 26-2-2020, https://www.efce.info/efce_media/European-Roadmap-PI-p-531.pdf
[6]   Wang, J, 2018, et al. Chem. Eng. Process. Process Intensif. 127 (2018) 111–126
[7]   Good Manufacturing Practice Guide for Active Pharmaceutical Ingredients, sourced 26-2-2020, https://database.ich.org/sites/default/files/Q7_Guideline.pdf
[8]   ICH Q8(R2), Pharmaceutical Development, sourced 26-2-2020, https://database.ich.org/sites/default/files/Q8_R2_Guideline.pdf
[9]   ICH Q9, Quality Risk Management, sourced 26-2-2020 https://database.ich.org/sites/default/files/Q9_Guideline.pdf
[10]  ICH Q10, Pharmaceutical Quality System, sourced 26-2-2020, https://database.ich.org/sites/default/files/Q10_Guideline.pdf

Part D: **Educating PI – academic and industrial**

# 18 Process intensification education BSc courses

## 18.1 BSc chemical engineering curriculum

Many curricula at universities of applied science consist of courses that teach the fundamentals of chemical engineering before culminating in a single design course (capstone design project) in which the fundamental knowledge and skills are combined to (re)design a process.

More and more chemical engineering departments want to have a strong focus on sustainable development, in combination with process innovation. The chemical engineering department at Utrecht University of Applied Sciences, for instance, has the ambition to add a strong focus on sustainable process innovation to its program. Process intensification is seen as an important element of that innovation. In this chapter, we discuss our (Utrecht) approach towards this goal, the challenges, and the opportunities we perceive.

The chemical engineering program consists of 4 years. In the first year, students are introduced to concepts from chemistry, physics, statistics, and mathematics and receive their first training in the technological lab. The second year prepares the students for their internship, teaching them about (among other topics) reaction kinetics, separation techniques, industrial safety, fluid flow, and process control. After a 5-month internship in a suitable company, the students' practical and theoretical training is continued in the third year, with topics such as heat and mass transfer and reactor technology. A major part of the third-year program is a course in process design. In the fourth year, a graduation assignment in the form of a second internship concludes the BSc program.

In order to implement sustainable process innovation into the program, we have chosen not to set up a specific course on process intensification, but rather to introduce it as a topic which is studied and applied in various courses. This approach gives the department the opportunity to acquaint students with PI principles in many courses, while developing knowledge about a fundamental engineering topic.

For example, during year 1, the students develop skills and knowledge related to the PI sustainability principles such as the circular economy theory. Students are challenged to think about applying these principles in an existing chemical (or manufacturing) process. Creative answers are expected by challenging each other in group discussions and individual contributions to the problem; twenty-first century skills are developed in the context of a theoretical problem.

In year 2, PI as a concept is introduced in the course "distillation & extraction." While students mainly learn to design (on paper) continuous and batch distillation columns by calculation and graphical methods, a single lesson is devoted to the principles of PI: what will distillation look like in the future? What options exist to

https://doi.org/10.1515/9783110657357-018

combine distillation with other unit operations, and what other applications of PI (outside reactive distillation, e.g., the microreactor) exist? Three batch distillation columns at the laboratory give students the chance to try out different distillation concepts in practice; practical hands-on work is a main theme in this part of the curriculum. Here too, PI finds its place. For example, SpinPro technology is used as an alternative to a mixer in an extraction experiment [1].

In the last 2 years of the BSc curriculum, the students specialize in a certain topic of their own choice; this can be, for example, biotechnology, the energy transition, water purification or food technology. This choice manifests not only in the internship and graduation project, but also in a project that small groups of students complete with (and within, for 1–2 days a week) a company. For both internships and this project, the department is in contact with over 150 diverse companies, specialized in themes such as regional (and city) wastewater treatment, innovative reactor concepts, dairy, plant design, catalyst manufacture, plastic recycling, and biorefinery. In these hands-on projects, students get the opportunity to apply the newly learned PI principles on the job.

The most direct application of the tenets of process development from a process intensification perspective remains the process design "capstone" course in the third year. In this course, students design and develop a new factory for a certain product and apply PI principles to optimize their design.

Finally, in a special minor focused on process development in the chemical industry, PI principles are studied in a multidisciplinary way. The minor allows students from different backgrounds, such as chemistry, management, and mechanical engineering, to collaborate with our students on (e.g.) the design of a new process. When tackling this from a chemical perspective, students are inadvertently faced with challenges and questions in relation to logistics, business cases and safety, health, and environment issues.

## 18.2 Necessary basic infrastructure to study PI

To teach PI at a University of Applied Sciences the basic facilities needed are good access to the latest developments in PI knowledge such as PI teaching specialists, access to journals, conferences, and workshops. Teachers at the University of Applied Sciences, Utrecht are motivated to continuously develop knowledge in the field of applied science, including PI. Given these tools, teachers are able to develop a vision on PI and share the latest developments with their students, using modern didactical methods. Secondly, highly developed practical facilities in a dedicated chemical engineering laboratory with well-trained instructors are the key to properly teaching PI. For example, at the Applied University of Utrecht, we work in close collaboration with equipment companies such as Flowid and NX Filtration and use their materials in our practical courses. The combination of theory and practice is invaluable in allowing PI knowledge to sink in. By further allowing students to perform research in and with companies, the bonds between teachers, students and companies are strengthened, allowing for intensive collaboration that benefits all parties. As mentioned before, a design project, late in the curriculum, provides a perfect opportunity for students to apply their PI knowledge and experience its consequences on process efficiency, cost, and sustainability.

## 18.3 Didactical methods

For several years, views on didactical methods in BSc programs have been shifting from a teacher-oriented approach to a student-oriented one. Instead of the teacher as an authority, sharing his knowledge with the students, nowadays, the role of the teacher is seen as more of a guiding one, helping students to solve problems. More attention is given to problem solving, collaboration and studying independently (at home or at school). A strong intrinsic motivation and interest in technological problems is required for success. This gives enormous opportunities for teaching PI in a BSc setting. Studying together and working on real industrial or societal problems, leading to a sustainable design solution becomes the norm, rather than an exception. This book contains many examples that can be used as foundation for process design challenges for group work.

All previous paragraphs focus on a certain type of student: the full-time student who enrolls in the program after finishing high school. At the University of Applied Sciences, Utrecht, it is also possible to obtain a BSc degree while also working a full-time job in the chemical industry. Process operators intrinsically motivated by the topics in chemical engineering and willing to get a degree use this route to become theoretically trained at a BSc level. They can immediately apply their knowledge and skills on the job. This gives enormous potential to bridge the gap between academics and companies. Teaching this group of students about PI creates an opportunity to open their eyes, and those of their supervisors, to the benefits of PI. "How to convince your supervisor to try PI" seems a valuable skill to teach this group.

It can be concluded from the perspective of Universities of Applied Sciences, that the most suitable approach to PI education in this phase of the student's career is a practical bottom-up approach, in which students apply principles on existing applications: identifying the concepts, practicing them in the lab, and combining them into a process design makes a student familiar with the concepts and applications of PI and increases the likelihood that the student will propagate his knowledge in a later work setting. In addition, there is a strong synergy between PI and recent innovations in education.

Chapter 13–17 can be used for familiarization of concepts with practical cases. Functions and function integrations, for instance, can be identified. Input and output streams can be made and checked for mass balances and or atom balances.

## Reference

[1]    Flowid, SpinPro reactor, sourced 17–2, 2020, https://www.flowid.nl/spinpro-reactor/

# 19 Industrial sustainable intensified process innovation training

## 19.1 Introduction

### 19.1.1 Maarten Verkerk management experiences leading to this book

I have worked a long time as a manufacturing manager in Philips. I have managed small factories and large factories. All these factories were within the ceramic process industry. In the mid-nineties of the last century, I was invited to follow an intensive course on "World-class Manufacturing." The idea was that every factory had to develop itself to the level of world-class manufacturing. It was a course about changes in the market, logistics, quality approaches, social innovation, and world-class manufacturing techniques.

The whole course was planned in three different hierarchical levels. The first level was the board and business unit management, the second layer was plant management and factory management, and the last layer was lower management and operators. The hierarchical structure was also translated into a schedule: first board and business unit management, after that plant management and factory management, and finally lower management and operators. The main idea behind this course was that the whole company understood the need to reach the level of world-class manufacturing, mastered the same philosophies and techniques of manufacturing, and spoke the same language. It goes without saying, world-class manufacturing became one of the main items during management reviews.

I followed the course on plant management and factory management. The whole course consisted out of three blocks of one week. These blocks week planned in a period of three months. After that the plant managers and factory managers had to train their people on the concept of world-class manufacturing. From my perspective, it was a great course. The philosophies of world-class manufacturing challenged my thinking. It required social innovation in traditional factories. Also, the different techniques of world class manufacturing were, to a large extent, new to me. In summary, this course changed my way of thinking and acting.

Why do I tell this story? In my opinion, the philosophies of world-class manufacturing and sustainable process intensification have a lot in common. First, they embody new ways of thinking. The German language has the beautiful word "umdenken." It expresses that you have to think in a different way – a way that is not compatible with the present thinking. Second, they embody new methods. The philosophy of world-class manufacturing offers a lot of new methods to reach this goal. The philosophy of sustainable process intensification also offers a lot of methods to realize its goals. The methods have been described in Part A of this book and have been applied in part B.

https://doi.org/10.1515/9783110657357-019

Finally, the philosophies of world-class manufacturing and sustainable process intensification can strengthen each other. Or in other words, sustainable process intensification is the next step to take. Training this next step is described in the following sections.

### 19.1.2 Jan Harmsen engineering experiences leading to this book

In my chemical engineering study at Twente University, I became interested in the problem of process scale-up. Somebody explained to me that the sole purpose of the chemical engineer is to solve the problem of scale-up from the lab experiments by the chemist to the commercial-scale implementation. However, in the chemical engineering education, little attention was paid to this scale-up, There was only an elective course: "Scale-up" based on dimensionless numbers. In 1977, I joined Shell and started in the department of Equipment Engineering at Shell Technology Centre, Amsterdam. In that department, more than one hundred engineers worked on scale-up problems for new processes. Here, I learned that the scale-up of new processes was an enormous problem involving design, modeling, and experimental validation.

In 1981, I started to work in Shell Research Laboratory, Sittingbourne, on biotechnology processes. I met John Linton, a microbiologist, who had discovered a micro-organism that emits protease enzymes through its cell wall. The enzyme can then simply be obtained by micro filtering off the micro-organism from the water, leaving the enzyme in the water. In commercial-scale production, the micro-organism of protease enzyme did not excrete its enzyme, and costly separation steps were needed to obtain the enzyme. This discovery, however, was not commercialized within Shell. The reason was that Shell had no experience in biotechnology processes and no business in enzymes. John was allowed to explain his discovery to Stirch, a fermentation firm, who immediately applied the recipe in their fermenters and made a profit in selling the protease enzyme. This story taught me that innovation has to be linked to the company business.

In 1984, I went back to Shell, Amsterdam to work on an oil-from-shale process. Heinz Voetter, a very experienced chemical engineer, had designed a novel process based on fluidization, which appeared to be superior to all known oil-from-shale processes, for two reasons. First, the process design was for a single train capacity 80,000ton/day of shale, while all other processes had a single train maximum of 4,000 ton/day. Second, the yield of oil on shale was also superior. The remarkable thing was that Voetter had made the design prior to any experiment. This taught me that process design, very early in the discovery stage can be an enormous advantage.

Hence, I focused once more on reading publications on design methods and found publications by Jeff Siirola on task-based process concept design, which he called "process synthesis." I became interested in process intensification as a new design method. In 2006, I invited Jeff to Shell, Amsterdam for a one-day lecture on his design method. He talked all day, and everybody was impressed with him and his design method. I kept in contact with Jeff, resulting, amongst others, in the Eastman case of Chapter 14.

When I left Shell in 2010, I started my consultancy firm, wrote a book on process scale-up [1], and provided courses and advice on process scale-up. In this way I came into contact with several smaller and larger companies, all struggling with process innovation. I did not fully understand why they had this problem.

Then, I met Maarten Verkerk, because I became his curator for his professor chair philosophy of technology in Technical University Eindhoven and Maastricht University. He invited me to be the coauthor of a chapter on industrial innovation practices. At first instance, I could not see any benefit in this cooperation, but Maarten kept talking and explained the importance of his approach. However, I remained skeptical.

Half-a-year later, Maarten phoned me again and told me that his chapter about industrial innovation practices was accepted by the editors of the book. He had written that chapter together with a former student of his, a process engineer. He offered me to send a copy and to discuss the content of the chapter over a glass of beer. So, we had a glass of beer, discussed his model – the so-called Triple I practice model – and explored the business opportunities. Maarten, again, was very passionate about his model and could not understand that I did not share his enthusiasm. I was still skeptical. I thought: "May be, it will have some benefits, but there is no evidence that this model applies for the process industry. What about an experiment with him to see what happens?" So, I told him about my plan to write a book about process intensification and I offered him to be a coauthor for the "soft parts" of the book. He immediately accepted my offer.

He wrote a revised version of his original model for process intensification. He renamed his model: VIB: values of engineers, interests of stakeholders, and beliefs of society. After I had read his chapter, we had a tele-conversation. I said: "Maarten, your text is a set of statements with no proof. Please provide me with some evidence from your work in the ceramic processing or from your work in healthcare. And I still also want proof from the process industry sector. How do we get some proof that the model fits to process intensification?" Maarten said: "Why don't you interview the writers of the industrial cases about the VIB items?" So, I did. And then a remarkable thing happened. All industry people immediately started to talk and write about the VIB aspects in their innovations. It was as if I had pressed a button and the flood gates opened. The results of that are found in Chapter cases 13–17.

Maarten then rewrote his part and I started to interview the industrial case writers in detail by using the VIB perspectives. All of them immediately started to write and to talk on it.

A prime example of what happened in these interviews is the interview with Jeff Siirola. I had arranged to meet him at the AIChE Fall meeting in the United States. I interviewed him about the Eastman Methyl Acetate process intensification case. To my surprise, he liked the VIB questions very much and answered all of them without any hesitation. He was very happy to tell the "soft side" of the methyl acetate story. It was as if for the first time in his life he could tell the whole story. He offered a detailed insight on the values of the engineers (V), the importance of identifying the interests of stakeholders (I), and the beliefs in society (B). He also spoke about the importance of education and the necessity of multi-disciplinary co-operation. In other words, he told the whole story of a successful process-intensification project. The result is found in Chapter 14.

The interview with Jeff Siirola was very insightful for me. First, he confirmed the idea that the VIB model was not a "theoretical model" but that it was a model that reflected the "flesh and blood" of the industrial engineering practice. Second, it confronted me with my own view about technology. I always focused on the "hard" site of the process industry (design, models, experimental validation) but I ignored the "soft" side of it (values, feelings, interests, basic beliefs).

May be, some of the readers of this book will have the same experiences when they read Chapters 3, 4, and 5 about the "soft" side of technology innovation. May be, their first reaction is "it makes no sense." But I do advise them: Read the industrial cases Chapters 13–17, then continue reading the theoretical chapters, and then the penny may drop.

### 19.1.3 Learning points

The stories of Maarten and Jan offer us a lot of learning points. The first learning point is that the introduction of new concepts like process intensification is not "just" a technological change but always involves a (radical) cultural change. It cannot be taken for granted that individuals adapt themselves easily to this change. That needs training and dialogues.

The second learning point is that in case of new concepts like process intensification, training needs to be done in a cascaded way. It starts at the top, and from there, the whole organization has to be trained. May be, the training also has to be extended to key stakeholders.

## 19.2 Need for industrial training

The lesson from the above stories is that the introduction of sustainable process intensification innovation is about two worlds meeting each other – The world of management and the world of engineers. For successful innovations these worlds meet each other, as is shown in the successful industrial cases in Chapters 13–17.

Training is, in general, the way to organize this meeting of worlds. This means all management levels are to be trained in this subject: top management, business unit management, plant management, factory management, and lower management; and industrial engineers have to be trained in sustainable process intensification innovation: both the hard design aspects and the soft VIB aspects.

Additionally, the company has to reflect on its networks crossing stage boundaries and crossing departments of R&D, marketing and sales and operations. There is no single recipe to develop these fruitful networks. However, the cases in this book suggest that the combination of passionate engineers and receptive managers could be the key to realize these innovative breakthroughs.

The need for management and industrial engineers' training cannot be underestimated. When, I (Maarten) started my job as a factory manager in an organization that was trained in manufacturing excellence concepts, I took it as a starting point that my management team and the industrial engineers knew the concepts and understood its philosophies. The main reason for this supposition was that they spoke the language.

After a couple of months, however, I discovered some resistance to these concepts and philosophies in my management team and my engineers. At that time, I interpreted this resistance as a sign that the implementation of the new ideas was a tough job. It took nearly a year for me to realize that my managers and engineers used the words that are characteristic of world-class manufacturing, but they did not understand their meaning in detail. Additionally, they used the methods, but they used them "too technically." I realized how important training and retraining are. The meaning of a new philosophy and its techniques will be understood well only if every manager and every engineer struggles with it. On top of that, it is not about the understanding of one "single" manager or one "single" engineer, it is about the understanding of the whole practice: R&D practice, OPS practice and M&S practice, The new philosophy and its techniques have to become a part of the DNA of the industrial innovation practice.

## 19.3 Management training

The Philips story discussed in Section 19.1.1 offers an example of how management training in green or sustainable process intensification can be done. Basically, it is a cascaded training; all hierarchical levels receive the same training: the philosophy and the techniques of green or sustainable process intensification. However, their focus is different. The broad objectives of the top management cascade down to the lower levels and will be defined more specifically.

First, the training of board and business unit management focuses on the changes in the business environment and its consequences for the company. It is about a new vision and new philosophy for the company. It is about repositioning the company. Second, it is about short-term and long-term strategies on how to realize these visions and philosophies. May be, the most important part of this strategy is to focus on industrial innovation practices and to develop new innovation networks. Third, it is about management of change. Not a "simple" change but a paradigmatic change.

The training of plant managers and factory managers also covers all elements present in the training of board and business unit management. On the one hand, this group of managers has to contribute to the new vision and philosophy of the company by translating it to their own plant and factory. It goes without saying that a catch-ball process with the board and business unit management is required to enrich the vision and philosophy and to make it happen. On the other hand, the strategy to renew the industrial innovation practices, to develop new innovation networks, and to implement the paradigmatic changes in the company has to be addressed extensively.

Preferably, the training of lower management and operators has to be organized by the local plants. The new vision and new philosophies have to be sketched and discussed. The consequences for the business unit and the plant have to be discussed extensively. The focus of the training will be that part of the techniques that are relevant for lower management and operators. It goes without saying that the role of plant managers and factory managers is not to be limited to opening and closing the course. They have to form its backbone .

It should be noted that the focus on process intensification, SDGs, and the VIB model has consequences for the investment policies, cost-benefit analyses, and payback times.

## 19.4 Industrial engineers training

The Eastman case of Chapter 14 clearly shows the importance of the training of engineers. Jeff Siirola sent his engineers to a process-intensification course at the university to make sure that they were familiar with the principles of process intensification, were trained in its techniques, and spoke the same language. All cases, however, show that there is far more required. Values of engineers drive the breakthrough innovation against all odds. Taking the interests of all stakeholders and values of society into account facilitates acceptance and finding funds. Multidisciplinary cooperation deters projects from halting half-way or resulting in a commercial implementation disaster.

First, in the training for engineers, it is of utmost importance to address the vision and philosophies of sustainable process intensification. They have to understand the real meaning of this vision and philosophy. Second, they have to be trained extensively in the principles of process intensification as discussed in Chapter 2. As far as possible, this course has to not only address general principles, but also has to apply these principles to the chemical processes of the company. Third, engineers have to be trained in the techniques presented in the Chapters 3–5. In particular, they have to reflect on their own values in relationship, the values of the organization, and the values inherent to sustainable process intensification. Finally, they have to be familiar with the different principles of management of change.

We argued in Chapter 5 that the practices of industrial engineers in R&D, Operations, (OPS), and Marketing & Sales (M&S) are quite different. They all have to understand the changing business environment, the vision and philosophies of green or sustainable process intensification, and its implications for the organization. However, their foci will be different. R&D have to be focused on the development of intensified processed, OPS on the control of these processes, and M&S on the business aspects.

## 19.5 Stakeholders training

We have argued in this book that Process Intensification requires a breakthrough in industrial innovation practices and industrial infrastructures and networks. For that reason, it has to be considered that the training has to be extended to the most important stakeholders.

## Reference

[1]    Harmsen, J, Industrial process scale-up – a practical innovation guide from idea to commercial implementation, 2nd revised edition, Elsevier, Amsterdam, 2019.

# Appendix 1  Pharma case: production increase by combined evaporation crystallization and cooling crystallization

## A.1 PI process

This process intensification case is related to a production increase of the final API crystallization to double the production capacity without increasing the reactor nominal volumes. It is the crystallization of an organic HCl salt where yield and production batch capacity are maximized without changing the crystallization solvent system, by combining an intelligent solvent switch: evaporation and cooling crystallization. The product quality concerning impurities as well as crystalline quality parameters are unchanged and the quality control is built into the process design. The critical objectives in process changes are to maintain product quality and to minimize regulatory work, which eventually means shorter processing times for getting Authority approvals for changes. In this final API crystallization system two solvents, methanol (boiling point 65 °C) and acetone (boiling point 56 °C), were used due to acetone-specific solubility of impurities and because of product solubility to methanol. Two reactors were used and the reactor sizes were unchanged. The charged batch size could be increased from 660 to 1,045 kg. The product yield was increased around 20%, and the processing capacity was increased 1.9 times compared to the old process.

The final API crystallization was the same as described in Figure A.1, except the addition of methanol distillation and the addition of acetone solvent. Details are found in Figure 17.3. In the first dissolution reactor, the crude product is charged along with methanol. The product is dissolved in the heated process solution. The undissolved impurities and foreign matter are removed during liquid filtration.

The starting point for this PI case was to study the old process development reports. Impurity and product solubility data was developed from the beginning of the R&D work back to the 1980s which gave the idea to use only methanol in the dissolution of crude product for final crystallization. However, the acetone addition was included to target minor changes in change control and also to maximize impurity removal efficiency. Addition of acetone, after methanol evaporation, was obvious because of solvent boiling points. It was clear that the focus was on product yield and capacity increase, as this PI task was initiated by the larger in-house project of re-evaluating chemical processes. Crystallization metastable region data demonstrated yield improvements in lab-scale and therefore larger solvent amounts of methanol were used in the dissolution reactor. An obvious unit operation was methanol evaporation before acetone addition, seeding, and cooling crystallization.

The methanol solvent removal by distillation is performed after hot liquid filtration in the crystallizer reactor. Distilled methanol is circulated via solvent regeneration.

https://doi.org/10.1515/9783110657357-020

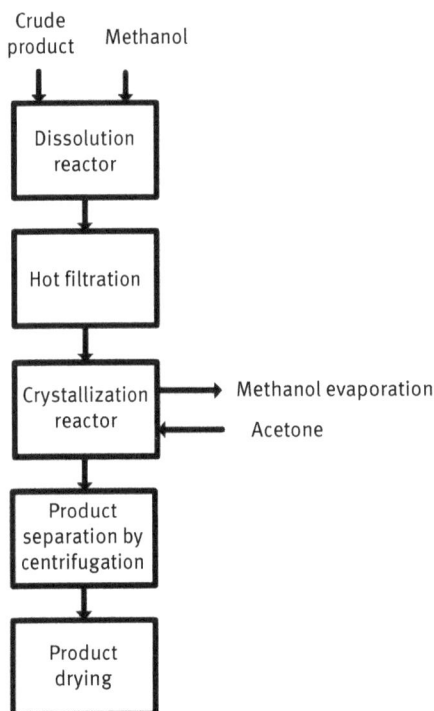

**Figure A.1:** Block diagram of unit operations in combined evaporation crystallization and cooling crystallization.

The addition of acetone amount is selected based on the same methanol/acetone ratio as in the old crystallization process. The API solution concentration was increased from 14 wt% to 33 wt%. The product is crystallized from solution by seeding and cooling the solution to end temperature, followed by centrifugation, drying, homogenization, and product packaging.

The old process was performed by charging crude product, acetone, and methanol to dissolution tank. The methanol/acetone ratio was 0.57. The heated solution was filtrated to remove foreign matter. The process was followed by seeding and cooling crystallization, centrifugation, drying, homogenization, and product packaging.

The profitability improvement of the new project, over the existing practice of processing, was based on improved processing capacity and product yield as a result from increasing the API solution concentration and as a result of evaporating methanol in the crystallization reactor. The capacity could be increased by a factor of 1.9 and yield could be increased by 20% without changing the reactor sizes. No obstacles were found for the PI project feasibility because the final solvent ratios, seeding policy, and cooling profile in crystallization was within the limits of the API Filing material. In centrifugation, the solvent washing amounts were also proportional to those in the old process and the final product was dried without changing the drying parameters.

As summary, the process steps comparison between the old and intensified processes was collected as shown in Table A.1. The values of engineers and interests of stakeholders (environmental permits from authorities, QA&QC, and eventually company ownership) are also summed up in that table.

**Table A.1:** Comparison between old process and Process Intensified process: Production increase by combined evaporation crystallization and cooling crystallization.

|  | Old process | New process |
|---|---|---|
| Solvent charging | Acetone and methanol into dissolution reactor | Methanol into dissolution reactor |
| Impurity removal | Hot solution filtration to remove foreign matter | Hot solution filtration to remove foreign matter and insoluble impurities |
| Solvent switch | All solvents charged at once | 70% from charged methanol distilled, acetone added after distillation |
| MeOH/acetone ratio in crystallization, | 0.57 | 0.57 |
| API concentration in crystallization, wt% | 14 | 33 |
| Work-up | Centrifugation, washing, drying | Centrifugation, washing, drying |
| Yield, % | 75 | 90 |
| Product batch capacity, kg | 500 | 940 |
| PI concept | Multifunctions in reactor 1 and hot liquid filtration: (1) addition of product dissolving solvent, dissolution and efficient impurity removal due to dissolving product, and not main part of insoluble impurities. Functions in reactor 2: (2) solvent evaporation, impurity dissolving solvent addition, cooling crystallization | |
| Value of engineers | Idea discovery by conducting experimental plan, concept and feasibility tested by simulating industrial-scale processing steps at a lab, development and implementation included by scaling up and process validation, NO PROCUREMENT OR EQUIPMENT CONSTRUCTION | |
| Interest of stakeholders | Owner's perspective: Intermediate product price reduction. Customer perspective: maintaining same product (impurity) specifications. QA/QC department: maintaining same product (impurity) specifications. | |

The concept was verified in laboratory experiments and as continuation bench scale batches were performed. In this specific case, the development was straight-forward based on laboratory scale development. The process profitability was easy to evaluate even based on lab experimentation.

The evaluation is summarized in Table A.1. The main learning point was that the solubilities of impurities and product API because of the addition of the evaporation step and the process operation order could be clarified. The reactor volumes were unchanged due to those process modifications.

## A.2 Multidisciplinary cooperation

The cooperation of the company's process chemist during the project was essential, first, to get the approval from head of R&D to go with PI idea, and then during the transfer to actual production. Analysis methods with specifications were available in this case and the process equipment in the factory plant were the same as in the old process. However, the suitability of the new processing methods were evaluated. The final process change evaluation was made before going to industrial batches, where responsible personnel from the factory, including HSE, QA and registration were involved. The process details were discussed with the representatives from plant operations and finally training sessions were held along with detailed processing instructions.

# Index

https://doi.org/10.1515/9783110657357-021